# Dialogues on Free Will

What kinds of conversations might take place when two intelligent and informed young people – one a strong believer in free will and the other just as certain that we don't have it – get together to discuss their viewpoints? This new volume in the *Philosophical Dialogues on Contemporary Problems* series introduces both sides of this age-old philosophical debate in a fun and accessible way.

Author and acclaimed philosopher Laura W. Ekstrom has created Rosa and Leif, two college students, who each give voice to the strongest arguments in the most important debates about free will and moral responsibility. Over the course of eleven days, Rosa, a strong defender of free will, draws on her religious beliefs and her thoughts about love and meaning in human life. Leif, her equally informed and articulate interlocutor, raises challenges to Rosa's points, drawing on the latest research and ideas in neuroscience, theology, and social psychology. Along the way, they must carefully develop clear reasons for their positions, while offering a fair hearing to alternative viewpoints and spurring readers to reach their own conclusions.

Working in a philosophical tradition established by Plato, these dialogues are both philosophically substantive and provocative, inviting readers to ponder issues that are fundamental to our understanding of the nature of choice, responsibility, and ourselves as human beings.

**Laura W. Ekstrom** is Chancellor Professor of Philosophy at William & Mary. She is the author of *Evil and Theodicy* (2023), *God, Suffering, and the Value of Free Will* (2021), and *Free Will: A Philosophical Study* (2000), as well as numerous academic articles.

**John Martin Fischer** is author of the *Foreword* and distinguished Professor of Philosophy, University of California, Riverside, and University Professor at the University of California.

## Philosophical Dialogues on Contemporary Problems

*Philosophical Dialogues on Contemporary Problems* uses a well-known form – at least as old as Socrates and his interlocutors – to deepen understanding of a range of today's widely deliberated issues. Each volume includes an open dialogue between two or more fictional characters as they discuss and debate the empirical data and philosophical ideas underlying a problem in contemporary society. Students and other readers gain valuable, multiple perspectives on the problem at hand.

Each volume includes a foreword by a well-known philosopher, topic markers in the page margins, and an annotated bibliography.

Dialogues on Gun Control
*David DeGrazia*

Dialogues on Human Enhancement
*Nicholas Agar*

Dialogues on Democracy
*Julia Maskivker*

Dialogues on Ethical Vegetarianism, 2nd Edition
*Michael Huemer*

Dialogues on Immigration and the Open Society
*Chandran Kukathas*

Dialogues on Free Will
*Laura W. Ekstrom*

For more information about this series, please visit: www.routledge.com/Philosophical-Dialogues-on-Contemporary-Problems/book-series/PDCP

# Dialogues on Free Will

Laura W. Ekstrom

Routledge
Taylor & Francis Group
NEW YORK AND LONDON

Designed cover image: Getty Images / eyecrave productions

First published 2026
by Routledge
605 Third Avenue, New York, NY 10158

and by Routledge
4 Park Square, Milton Park, Abingdon, Oxon, OX14 4RN

*Routledge is an imprint of the Taylor & Francis Group, an informa business*



ISBN: 978-1-041-16376-3 (hbk)
ISBN: 978-1-041-09663-4 (pbk)
ISBN: 978-1-003-68426-8 (ebk)

DOI: 10.4324/9781003684268

Typeset in Sabon
by Newgen Publishing UK

# Contents

# Foreword

John Martin Fischer

The topics of free will and moral responsibility have been of central interest to philosophers, theologians, scientists, and thoughtful persons for centuries and even millennia. This is in part because they are intertwined with questions about our very humanity – our identities as persons; our most fundamental relationships (love, friendship, business and athletic collaboration and teamwork, and so many more); the legitimacy of moral praise, reward, blame, and punishment; and our relationship to God (if there is a such a higher power). This list just scratches the surface.

The topics have generated a huge body of discussion and debate over the years. It is striking that the intense interest in these topics continues to this day, with lively disputes in the academic journals and scholarly books, and venues for broader public discussion. The focal points of the debates have changed somewhat, although not entirely, over the centuries (and even decades), but some of the fundamental issues remain the same. Current thought about free will and moral responsibility, unsurprisingly, has begun to explore the myriad relationships between the traditional treatments of these topics and recent (and accelerating) developments in artificial intelligence (AI) and neuroscience.

All this is fascinating and exciting! It is both intellectually deep and important in various applications. Because of the huge body of literature and complex historical and contemporary debates, however, it is challenging for a beginner (or even someone who has grappled with these issues for a long time) to grasp the overall picture and, certainly, to keep up with current developments. This is why the book you are about to dive into – Laura Ekstrom's contribution to Routledge's *Philosophical Dialogues on Contemporary Problems* series, *Dialogues on Free Will* – is such an invaluable tool.

The book is remarkable in many ways, a few of which I will note. It is comprehensive, tackling almost all, if not all, the many issues pertaining to free will and moral responsibility. Given this large (and treacherous) terrain, the broad coverage is no small accomplishment. Because the great historical debates illuminate the contemporary discussions, the book delves into the traditional contributions of philosophers, theologians, and other thinkers. Ekstrom both highlights the insights that emerge from the swath of historical discussions and shows how they relate to, and shed light on, contemporary explorations of these issues.

The book is written in a clear style that is easily accessible to students and scholars at any level, as well as the general public. The dialogue style makes it both lively and "down-to-earth." The annotated bibliography is an extremely helpful addition, which provides guidance to both those who are first encountering a more formal treatment of these issues and many who are already familiar with some of it. This allows the reader to choose to pursue issues of special interest in greater depth. In short: the book is a highly engaging and helpful introduction to this area of academic, religious, and more general human inquiry. It is written by someone who is an active contributor to the contemporary philosophical debates and is thus well-versed in the voluminous literature. It helps the reader see *what* the issues are, *why* they are important, and *how* to think about them productively.

In his poem, *Paradise Lost*, John Milton described the fallen angels as debating free will forever and being "in endless mazes lost." (They evidently never ran out of topics to discuss!) The debates can be complex, and the various philosophical pathways hard to navigate, but Laura Ekstrom is an excellent guide.

*John Martin Fischer,*
*Distinguished Professor of Philosophy,*
*University of California, Riverside*
*University Professor in the University of California*

# Acknowledgments

Free will has fascinated me for decades, connected as it is to so many issues of enduring human significance: meaning in life, the nature of our selves, what it means to be agents, human suffering, the existence of God, causation, determination, and moral responsibility. Through the years I have benefitted from conversations on these topics with many philosophers, including Susan Wolf, Nomy Arpaly, Dan Speak, Lilian O'Brien, Neal Tognazzini, Philip Swenson, Joe Campbell, Mylan Engel, Keith Lehrer, Robert Kane, Al Mele, David Robb, Alicia Finch, Brie Gertler, Derk Pereboom, Brian Leftow, Richard Swinburne, Dana Nelkin, Peter van Inwagen, Alvin Plantinga, Eleonore Stump, Manuel Vargas, John Martin Fischer, Mark Balaguer, Saul Smilansky, Tom Senor, and Michael Bratman. It was Michael Bratman who first introduced me to the topic of free will in his lively philosophy of mind course at Stanford. I have also had the pleasure for many years of teaching and conversing with wonderful students at William & Mary, who continue to impress and delight me with their active engagement in philosophical discussions. I am grateful to all of these people, as well as to an anonymous referee and to Jonah Goldwater and Dan Speak for very helpful comments on the manuscript. Thank you to series editor Andrew Beck for the invitation to write this book, to Phoebe Oliver for editorial assistance, and to the two anonymous reviewers who gave terrific input on the initial proposal. Finally, thank you to Kristen, Justin, and Matthew, for all you do and all you mean to me.

# Prologue

**Abstract**

*Two friends who are university students, Leif and Rosa, agree that they will meet regularly over the next couple of weeks to discuss their ideas about free will. They decide that, first, it will be Rosa's turn to present her reasons for thinking that we human beings do have free will. She remarks that she has not before now been prompted to say why she believes in free will, so initially she may have a hard time saying more than, "we just do." Rosa agrees that, after some reflection, she will take the lead for the first five days of conversations. After that, Leif will give his reasons, over the span of the next five days of conversations, explaining why he thinks that we do not have free will. Finally, the two will get together one more time to talk about how their thinking has changed over the course of their conversations. They will sum up what they have found they can agree on and identify where they still disagree.*

**Rosa:** Hi, Leif!

**Leif:** Hey, Rosa. Do you have a minute to talk about something?

**Rosa:** Sure.

**Leif:** My dad said something on the phone the other day that rubbed me the wrong way, and it's been on my mind.

**Rosa:** What was it?

**Leif:** He was talking about people who are unhoused, living in shelters or on the street, in connection with his political

DOI: 10.4324/9781003684268-1

views. It sounded like he assumed that having no home is a person's own fault and due to their own *free choices*. But that sounded off-base to me, because in several of my science and social science classes, I've learned that we don't really have free will.

**Rosa:** Yeah, I can see why that would be confusing and upsetting.

**Leif:** Thanks. It's been making me want to do more reading on the topic and to talk it over with someone.

**Rosa:** I get that. I'd enjoy having some more meaningful conversations outside of classes. I'm usually in the art studio working on projects, so I've become kind of anti-social lately.

**Leif:** Oh, me, too – except I'm always in labs. (Laughs.) So would you be up for talking about this in more depth sometime, I mean, about human free will?

**Rosa:** Sure, that sounds fun. Regarding people becoming unhoused, specifically, it seems to me like that's a complicated social, economic, and political issue, and I'm sure there are multiple causes that vary across different cases. Speaking more generally about free choice, though, yes, the topic of human freedom interests me, and I think obviously we *have* free will. I believe in it. Don't you?

**Leif:** Actually, I don't. I do not think we have free will. I definitely don't think it's obvious that we have it – no offense!

**Rosa:** (Laughs.) None taken. Okay, sure. It would be fun to figure out why we disagree. How do you want to go about it?

**Leif:** Well, what if you go first. We could get together kind of regularly for a little while, maybe over the next couple of weeks, if we could make time for it…how about we break up our discussions, so they don't run on too long for either of us to do our other stuff. You could start, giving me reasons why you believe in free will. Maybe you could elaborate on one reason each time we get together. Then when you're done, after several days of explaining your side, I could go. I'd try to explain why I think we do not have free will.

**Rosa:** That sounds like a good plan. The thing is, if I'm first…I don't really know where to start. No one has ever asked me before *why* I believe in free will. So it's hard for me to say more about our having it than, "we just do!"

**Leif:** Do you want to both think about it some more tonight, and then get together tomorrow afternoon to start?

**Rosa:** Sure, sounds good! I'm free late afternoon. Get ready, I'm going to prove you wrong.

**Leif:** Ha, I don't think so, but see you tomorrow. How about right here, around 4 p.m.?

**Rosa:** See you then!

# Part I: The Case for Free Will

# Day 1: Does Religion Show that We Have Free Will?


**Abstract**

*As Rosa gathers her thoughts, she begins in backing up her claim that we have free will by appealing to her religious background. Her initial assertion is that we have free will because God judges us based on our actions. When Leif objects that she is assuming that God exists, Rosa agrees, but says that she was taught that God exists by her religious upbringing. Leif acknowledges that lots of people have that experience in their upbringing and cultural background, so he agrees to explore her thoughts further. They clarify that their discussion takes place with respect to certain versions of monotheism, primarily in the Abrahamic traditions, according to which a perfect divine being holds human beings accountable, and our position in an afterlife is settled at least in part by the ways we freely made our decisions during earthly life. Their conversation covers the problem of evil, and they explore Rosa's appeal to human free will in response to this problem.*


**Rosa:** Hey, Leif! This is really nice, sitting on the grass under a blue sky in fresh air and sunshine. Okay, I've thought about it. I have a reason to give you.

**Leif:** I'm all ears.

**Rosa:** I think we have free will because God judges us based on our actions.

**Leif:** But, Rosa, you're assuming God exists.


DOI: 10.4324/9781003684268-3

**Rosa:** Yes, I am assuming God exists. I was taught that by my family, and I know that it's true.

**Leif:** Well, I'm aware that lots of people have that experience in their upbringing and family culture – that's very common. Still, we can't say we know that free will exists because we know that God exists, can we?

**Rosa:** I think we can. I'm certain that God exists.

**Leif:** That opens a whole other line of discussion. It's not certain to everyone that God exists, and people disagree over what God is like. Also, there are some pretty compelling reasons for thinking that God does not exist. The biggest of those, I think, is the huge amount of suffering we see in our world, which sure makes it look like there is not a loving, omniscient, all-powerful divine being who created this place and who is in charge of the universe. We see cancer, war, genocide, abuse… just so much pain and evil. But even if God were to exist, you are assuming that God judges us and that God judges us fairly only if we have free will. Isn't that right?

**Rosa:** Yes, that's what I think. The Scripture I believe in makes clear that God sees us, all the time, and judges us for what we choose to do, and even for what we think about. We end up in heaven or hell after our lives on Earth, based on what free decisions we make.

**Leif:** Okay, well, let's come back later to the idea of being judged for our thoughts. You're saying that you think that people who make good decisions, as judged by God, end up in heaven, and those who make bad decisions end up in hell. What is it, a measuring scale? Do you have to do more good things with your free will, even though you can do some bad things, but if you do more bad than good, you go to hell?

**Rosa:** Actually, in my specific Christian tradition, I think it's more a matter of having faith instead of rejecting God's offer of salvation.

**Leif:** And you think that's a free choice, to have faith or not, a choice people are responsible to God for making?

**Rosa:** Yes, I do.

**Leif:** I suppose how I'd think about that depends on what you mean by "faith." Is it a belief, or a conviction of a certain kind of strength? It doesn't seem to me like most of our beliefs are ones we freely choose to have or not to have…like my belief that you're sitting right there next to me. I just find myself with that belief. I see you, and I hear you talking, and I can't help believing that you are there and that you're not a figment of my imagination. But it's different with the case of God. I don't see God, and I don't hear God. It doesn't seem like I could just choose to start believing that God exists if I wanted to, especially since it seems to me that there are reasons for believing that there is no such being, really. It is just a figment of the human imagination, I think, and an invention to control people's behavior. So…do you mean faith as an emotion? Or as a belief you hold despite not being able to see God as obviously as you see me sitting in front of you?

**Rosa:** The faith I'm talking about as a matter of choice, in my religion, is freely accepting God's sacrifice on the cross, when Jesus of Nazareth accepted the punishment for our sins in our place. If we freely reject that good news of the gospel, we go to hell, because God's just judgment of our human sinfulness requires it, but if we freely accept Jesus into our lives as our Lord and Savior, and have a relationship with him, then we are forgiven for our sinfulness, and we can live forever with God in heaven.

**Leif:** There's a lot to unpack in what you just said.

**Rosa:** Yes, I know. My basic point is that I think we have free will because the collection of religious beliefs I hold includes our being responsible to, and judged by, God.

And since God is fair and good, God would not hold us responsible or judge us if we did not have free will. So, we must have it.

**Leif:** What do you say about people who believe that the right way to live – and the way to get to a heavenly afterlife with God – is to follow the teachings of the prophet Mohammed? Or other people who think that the path to God is by living by the tenets of one version or another of Judaism? Or what about people who are Hindu, or those who are Buddhist? Other people follow Jainism, or Sikhism, or the Bahá'í Faith, or Wicca, or Zoroastrianism, or Taoism…There are so many religious faiths and traditions in the world.

**Rosa:** That's true. But I think getting into that might take our discussion of free will too far off track. What I want to focus on is the billions of us, including myself, who believe in a kind of monotheism that affirms the existence of one God who judges us based on our own free choices. That's a very widespread belief, and it's one that I hold to fervently and use as a basis for lots of my other beliefs. In fact, I have some data right here that I looked up last night: a global study by the Pew Research Foundation (2012) found that more than 8 in 10 people identify as religious, amounting to 5.8 billion religious believers worldwide, including 2.2 billion Christians, 1.6 billion Muslims, 1 billion Hindus, and 14 million Jews, among others.

**Leif:** Okay, and your position is that the belief in monotheism – and I agree it is a very widespread position, even though there are many variations and interpretations of monotheism – serves for you, and for many other people, as a reason for thinking that we do have free will.

**Rosa:** That's right.

**Leif:** So, then, I guess we could take our conversation in the direction of debating whether or not God exists. Or, instead, we could talk about whether or not God's existence implies that we have free will.

**Rosa:** Let's not get too far off our topic by debating God's existence. I mean, just briefly, I can say that I think the universe needs an ultimate cause and explanation, which indicates that God exists. And I think that the universe shows evidence of design by an intelligent supernatural being, because it's so beautiful and intricate and amazing, with parts that work together to achieve goals, and it contains beings like us who are conscious, rational, moral people. Also, I don't talk about this often, but I have had some religious experiences, nothing like hearing a booming divine voice or seeing God directly in a bright light, but I've felt God's presence while I'm praying and worshiping – those experiences make me feel sure that God is real.

**Leif:** Okay, I agree – let's not get into debating the cosmological argument and the teleological argument for God's existence. And I don't want to offend you by getting into questioning your religious experiences, maybe by offering alternative explanations of them other than that God is really there. I want to stay friends! Let's try to stick to free will.

**Rosa:** Sure. I know not everyone has religious experiences, and I agree with something you alluded to earlier, that all the suffering that people and non-human animals go through on Earth does naturally lead to questions about how God could exist and why God would allow those things to happen.

**Leif:** I'm glad you've brought that back up. So, I want to try to recap your position regarding religion and free will. I asked why you believe in free will, and you cited your religious beliefs. So maybe this is your argument: first idea, God exists; second idea, if God exists, then we have free will; and therefore, conclusion, we have free will. Is that right?

**Rosa:** Yes, that sounds about right.

**Leif:** Let's focus on that second idea, rather than debating God's existence. You said part of your set of beliefs in God includes the belief that God judges us for our actions, and you think that we must have free will because of that fact. That's one of your reasons for thinking that the second claim – if God exists, then we have free will – is true. I was wondering if you also support that second claim by the belief that we must have free will in order for there to be a way to explain the evil actions and suffering in the world.

**Rosa:** Yes! Genocides, and betrayals, and racial injustice, and cruelty toward people and animals are not God's fault – they are our fault! People have the gift of free will, and they misuse it to cause pain and suffering.

**Leif:** So, your reasoning is that if God exists, then in light of the evils we see in our world, we must have free will, because God would not be the source of things like murder and racial hatred and cruelty to children and animals. Those must be "on us" – the results of our wrong free choices.

**Rosa:** That's right. And, as I said, I think free will was a gift God graciously bestowed on us. It's a way in which we are made in God's image. God did not create robots or puppets. God lets us make our own free choices about how we live, including the freedom to do really awful things.

**Leif:** So, let me pose a question for a minute here: why would free will be a gift?

**Rosa:** Of course it's a gift. Why would you question that?

**Leif:** Because if we have it, it's the source, as you just said, of genocide, murder, racial injustice, betrayal, cruelty... so many horrible things that people and non-human sentient beings suffer as victims – including violently-treated pets and animals slaughtered in factory farming. People have done so much that is evil, inflicting physical and emotional abuse on children that's virtually unspeakable...and they have tortured people and enslaved people. Why would a perfectly good Creator want to make human beings capable of doing things like that? Why would God create beings with the free will to do evil?

**Rosa:** Well, if we weren't free to do evil, then we wouldn't be free to do good, either.

**Leif:** Honestly, I don't see why not. What if we could just freely choose among a whole range of good actions, like, for example, in the next hour, being kind to my friend Kristen by helping her move some furniture, or instead cooking dinner for my family so my parents and siblings can do other things, or instead creating a gift for my grandmother?

**Rosa:** Would those even count as good actions, if there weren't evil for contrast?

**Leif:** Yes, they would – kind and helpful actions are kind and helpful even if there is no unkindness or narcissism.

**Rosa:** Wait, how can something be kind if there's nothing that's unkind?

**Leif:** Well, you said that you think God exists, right?

**Rosa:** Yes.

**Leif:** And I assume you think God exists timelessly or that God existed before and apart from the physical universe that God created?

**Rosa:** Yes.

**Leif:** Then you do think that good can exist without evil. Because you think that God existed when there was no evil at all, when it was just God in existence. And God is pure good, not at all evil, right?

**Rosa:** Yes, God can exist without there being any evil. So I guess that means good can exist without evil existing. But, still, I think none of us human beings would appreciate anyone acting in a way that is kind and helpful to us, if we didn't ever experience being treated in ways that aren't kind or in ways that are unhelpful, and if we never saw cruelty and mistreatment around us. Kindness wouldn't seem good to us without there being anything bad to compare it to.

**Leif:** Saying that we human beings wouldn't *appreciate* good things unless we also experienced bad things is different from saying that good cannot *exist* unless evil also exists. The point about our appreciation tendencies seems to me to be an observation about human psychology. I can see what you mean about our human tendencies – in that, for example, I appreciate my health more after I've gotten over being sick. And now that I have a knee injury, I appreciate that I used to be able to run up staircases without pain more than I appreciated it back when I could do it without pain! The thought of being in pain whenever I climbed stairs just didn't occur to me before.

**Rosa:** Right, that's what I mean.

**Leif:** But even if that is true as a tendency in human psychology – that we tend to appreciate good things when we lose them, or after we lose them and get them back, or when we compare them to worse situations – still, I think it's important to notice that we would not *have* to be made that way.

**Rosa:** Why do you say that?

**Leif:** I mean, if God is omnipotent, then God can do anything, right, at least within the bounds of logic. And it's not unthinkable or illogical that God could have made rational sentient beings like us but without that tendency you're referring to. God could make beings who do fully appreciate and enjoy goodness without a contrast.

**Rosa:** I don't see the point you're making.

**Leif:** God could create anything God wants to create, apart from things like square circles, since there is no such possible thing, a circular square being a contradiction. So, in creating physical beings like us who are intelligent and who feel sensations and emotions, God could give those physical beings a psychology that is full of joy and contentment and love and deep appreciation for goodness. There's no logical requirement for there to be strife in order for there to be joy. "Strifeless joy" is not a contradiction in terms like "circular square" is. A perfect God could create beings who experience pure joy and love all the time.

**Rosa:** I guess I can see that.

**Leif:** The existence of good, then, does not require the existence of evil. And appreciation for good on the part of created beings does not require that those created beings experience evil.

**Rosa:** It's still hard for me to grasp that second concept, because we do experience both good and evil, and we appreciate what is good in our lives in comparison to the evils in the world. I appreciate my having enough to eat every day, for example, especially when I think about people who are struggling in poverty or people who are starving in a famine. If I don't think about what it would be like to lack enough nourishing food, I don't appreciate it at all or at least as much.

**Leif:** I can relate to that. My point is just that God (if there were a God) would not *have* to make us like that. And the suffering of people who are starving to death cannot be justified by their helping us to appreciate our food! That would be cruel – God, as a wholly good being, wouldn't simply watch while some human beings suffer in the process of starving to death, *so that* others of us can appreciate our good fortune of not starving. That can't be how it works. It doesn't make sense.

**Rosa:** I don't think that God lets some people starve to death so that other people will appreciate not starving themselves. I think we can blame human beings for failing to distribute resources well so that no one in the world has to starve to death. Starvation has human causes, and we can prevent it. That's something we can use our free will to fix. So we are back to the topic of the gift of free will.

**Leif:** Let me recap – you've said we must have free will because God holds us responsible for what we choose to do, such as have faith in God or not, or live in a way that God wants us to live or not live that way. And you've said that we must have free will because it explains why there are evils in the world like starvation and murder and child abuse, even though God exists.

**Rosa:** That's right.

**Leif:** You would agree that a collection of religious beliefs would have to be logically consistent in order to be rationally defensible, wouldn't you?

**Rosa:** What do you mean by that?

**Leif:** I mean that your, and anyone's, collection of religious beliefs would need to fit together coherently – they'd need to mesh and not contain contradictions among them – in order for those beliefs to be ones you can defend and explain to other people, in a way that should prompt other people to see your set of beliefs as justified or rational.

**Rosa:** Yes, I'd agree to that. I think that my beliefs about what God is like, as well as my beliefs about human free will and human accountability to God, and my beliefs about why there's suffering in the world, these all fit together. I don't think there are contradictions between them. And I agree that, if there were contradictions, then my beliefs wouldn't be rational. But I think they are rational.

**Leif:** Okay, so let me go back to what you said earlier when you said that you believe that human beings have free will because you believe that God judges us for our actions and choices (and maybe even for our thoughts) – and as a result of that divine judgment, in the afterlife we end up either in heaven or in hell. I think there might be a contradiction in that whole picture, especially between believing in God and believing in hell as a kind of eternal punishment that could result from God's judgment of us. And if there is a contradiction there, then using your set of religious beliefs as support for the contention that we have free will is called into question.

**Rosa:** Why would you think that there's a contradiction between belief in God and belief in hell?

**Leif:** When I think of a perfectly good and perfectly loving God, I can't fit that idea coherently with the idea that some people suffer eternal torment away from God. I would think that God would love everyone, no matter what they did, and God would want everyone to live forever with their Creator who cherishes them. And anything God wants, given God's omnipotence, God can get.

**Rosa:** But God, as I conceive of God, is perfectly just, too – that is part of God's nature.

**Leif:** How could it be just for any people to suffer eternally for any choices they made here on Earth? Everything we do here is finite, even if it's really hurtful. How could any wrong finite actions merit punishment that goes on forever?

**Rosa:** Well, maybe hell doesn't go on forever. Maybe hell is punishment that is limited in time. I don't know. But let me ask you a question in return: how could it be just for people who rejected God their whole lives long – to take extreme cases, someone like an atheistic serial killer or someone who led a genocide and never repented, never asked God for forgiveness or salvation – to end up in heaven along with the people who spent their lives trusting and worshipping God and following God's commandments, treating other people while on Earth, as much as they could, with compassion and kindness?

**Leif:** I can see where, if you were a good and faithful person on Earth, and you found yourself in heaven next to an unrepentant hateful war criminal, you might think there's something wrong with this picture.

**Rosa:** Right! People who hate other people and reject God wouldn't want to be with God in heaven, anyway. They wouldn't choose it. So why would it make sense for them to be there?

**Leif:** Maybe because God loves them, no matter what they've done or how rejecting or cruel they've been, just like good and loving parents love their children even when their children behave badly.

**Rosa:** But suppose these parents you have in mind have two children: one of them meets his responsibilities – he completes his family chores, making his bed and taking out the trash, say – and their other child does not meet his responsibilities – he ignores the assigned cleaning tasks and acts completely selfishly. It wouldn't be fair for the parents to give them both the same reward, like ice cream as a treat, regardless of what the kids did.

**Leif:** But you mentioned that your faith includes the idea that people can behave badly and then repent, asking for forgiveness and accepting Jesus as their Savior, right?

**Rosa:** Yes, it does. So I suppose the parents could give both kids the same reward if the one who had been selfish was genuinely sorry, repented, and asked for forgiveness.

**Leif:** Okay, so you think there are some people who never do that, with regard to God and their own sinfulness, and since they don't, they end up apart from God in hell.

**Rosa:** That's right. What makes that fair is that we have the free will to decide to repent and ask for forgiveness or not, to turn toward God or turn away from God.

**Leif:** When you think of the loving parents we just talked about, don't you think they would continue to love their child who was selfish and who never apologized for not following the family rules?

**Rosa:** Yes, I think they would. Not all human parents are loving, but when I think of parents who are genuinely loving, I think of them as people who continue to love their children no matter what, unconditionally.

**Leif:** So, wouldn't that unconditional love imply that they would not allow any child they love to suffer horrendously in something like hell? They would do whatever they could to rescue their child from a hellish situation, to take care of them and keep them protected.

**Rosa:** I mean, if adult children cut off all contact with their parents and want nothing to do with them, the parents can't control them – they can't make their adult children love them or spend time with them. I think of it being that way with God and people who go to hell. People in hell choose not to be with God. And God respects their free choice.

**Leif:** Wait – human parents cannot control their adult children, that's true, but God's case is different: if God is omnipotent, then God can do anything that's logically possible, including bringing all human beings into a heavenly afterlife.

**Rosa:** Okay, well, God would not do that, because God gave human beings free will and respects our free choices, including the choice to reject God. God would not *override* our free choice and bring us to heaven against our will.

**Leif:** I don't see why we should think that. If I could keep my beloved adult child from committing suicide, I would do everything in my power to make that happen, regardless of their will. For someone to want to commit suicide, something must be very wrong, something like suffering a mental health problem such as clinical depression, and so I'd see my adult child as unwell and not in their right mind, and I would make every effort to get them protective care to keep them from suicide. I think God would do the same with respect to any supposed "choice" to go to hell, rather than to be in heaven with God and in the presence of full goodness and pure love and joy. No one in their right mind who is seeing the options clearly would choose that.

**Rosa:** I see.

**Leif:** To my way of thinking, then, if anyone were to end up in hell, it would not be their fault. They either didn't get the message accurately about what they were supposed to do or to believe, in order for them to get into heaven, or they chose on the basis of bad information or illness or confusion. A "free choice" to go to hell doesn't make any sense and, anyway, it would not be something God would choose not to interfere with or override, knowing the consequences for the beloved human creation, especially if those consequences involve intense suffering for eternity. That would be akin to my doing everything in my power to keep my (hypothetical) loved adult child from ending their life. God's power is not limited, and neither is God's love.

**Rosa:** I just have to say, again, that my religious tradition is clear about God holding us responsible for our choices and actions. God gives us clear commandments and instructions about how to live and how to follow God. We are responsible for learning these and obeying them by our own free will. And I want to reiterate how many people have religious beliefs like mine. Another study I found by the Pew Research Center (2021) looked at the views of U.S. adults regarding the afterlife: 73% said they believe in heaven, and 62% said they believe in hell.

**Leif:** That's by far the majority. Have we reached an impasse now?

**Rosa:** Well, I suppose it's possible that no one ends up in hell, apart from God, forever, because I can see your point that anything we do while on Earth is only finitely bad and maybe it does not merit infinite punishment. Still, maybe rejecting God, who has the most value it is possible for any being to have, results in eternal separation from God in hell, because we've sinned against someone infinitely valuable. I admit, I don't know exactly how it works or what to say about people who grow up in religious traditions other than mine. Still, I come back to the point that we must have free will because my religious beliefs include God's holding us responsible, and God's doing that fairly requires that we have free will so that our choices are our own, to our credit or discredit.

**Leif:** Let me approach this a different way before we leave the topic of your religious beliefs as your reason for believing in human free will. I've heard an interesting argument from the philosopher John Schellenberg which says that, if we have free will, then there must *not* be a God! Have you ever heard of that?

**Rosa:** No. What is the reasoning there?

**Leif:** Here's the idea. According to monotheism, the highest good for human beings is to be in a growing, learning, deepening relationship with God in which they share in God's love and presence.

**Rosa:** I would agree to that so far.

**Leif:** But then the key to this argument is this: for us to be in such a growing, sharing relationship with God – to achieve our highest good as human beings – does not require any free will of the kind that would allow us to do great evils.

**Rosa:** I don't see what you mean.

**Leif:** For God to create beings who are rational, who are capable of loving relationship with God, and who are capable of growing in knowledge of and closeness to God, it is not logically required that those beings have the free will to do evil things to other human beings or to non-human animals – things that are atrocious or that cause pain and suffering to themselves and others. Since our highest good does not require that kind of free will, *if* we actually do have that power of free will and it is the explanation for the evils of this world, then God must not exist. The reason is that God *would not have given us* such a power of free will, because of all the horrible things it brings about and because it is not required for our highest good. So, assuming that we have the kind of free will you've described, implies that God does not exist.

**Rosa:** Honestly, I don't know what to say about all that. I think I need to think about it some more.

**Leif:** Yeah, sure, of course. These are big topics, and I'm really enjoying our conversation so far. Want to get together again tomorrow, same time?

**Rosa:** Sure. See you tomorrow, Leif.

# Day 2: Does Interpersonal Moral Responsibility Show that We Have Free Will?


## Abstract

*After thinking more about what Leif has said the day before about God, suffering, and the highest human good, Rosa suggests that they leave God out of it, for the time being, and talk about human freedom apart from religious beliefs. Her next suggestion is that we have free will because people are responsible to each other for their choices. In response to Leif's question of what she means by that, Rosa discusses an example involving her long-term relationship with her boyfriend. This prompts Leif to talk about different theories about the central function of ascribing moral responsibility to people. He describes varying accounts of what is required for being morally responsible. What becomes clear is that the type of free will required for moral responsibility on these theories vary. This day's conversation helps Leif and Rosa to put these varying accounts on the table.*


**Rosa:** Hi, Leif. I've been thinking more about what you said yesterday about God, suffering, hell, and our highest good. For now, let's leave God out of it and talk about human freedom apart from religious beliefs.

**Leif:** Sure, good idea. You were right to say that many people – in fact, billions of people – are religious and that their religious ideas are often interwoven with their thoughts about free will. Not everyone, though, is religious or even cares about religious ideas.

DOI: 10.4324/9781003684268-4

**Rosa:** Right. So, moving on…Another reason I think we have free will is because people are *responsible to each other* for their choices.

**Leif:** What do you mean by that?

**Rosa:** Here's an example of what I have in mind. My boyfriend and I have been together for three years. We don't have an open relationship – we have promised to be exclusive, to date only each other. If he were to cheat on me, he'd potentially expose me to a sexually transmitted infection, and it would break my heart to find out because I trusted him, and we had promised to be faithful to each other. I think he would be responsible to me for making a choice to have sex with someone else. He would owe me an explanation and an apology, and he would deserve blame, and he would deserve to hear about my anger and hurt because of what he chose to do. And I'd be justified in breaking up with him if I chose to, for his violation of my trust.

**Leif:** I can understand that. So you're thinking that, as an example of the fact that people are morally responsible to each other for how they act, your boyfriend is morally responsible for how he treats you, including keeping the promises that are part of your relationship. And his being morally responsible for cheating on you, supposing he were to do that, implies that he has free will, at least with respect to cheating on you.

**Rosa:** That's right.

**Leif:** And could you explain what you mean, when you say that he has free will regarding cheating on you?

**Rosa:** I mean that, when he had an opportunity to cheat on me – the temptation at a party, or by way of a dating app, or whatever it was in the hypothetical situation – he could have either acted on that opportunity or instead walked away and chosen to stay faithful to me.

**Leif:** I see – he had options.

**Rosa:** That's right. Criminals, too, deserve to pay a penalty for what they freely choose to do wrong when they violate the law. They have options.

**Leif:** What if someone is hungry and steals so as not to starve? Are there really options there?

**Rosa:** They should ask for help from a shelter or a food bank, not steal. That's their free choice.

**Leif:** I'm curious about that, though...because what if someone was raised in an environment where people who steal are treated with admiration, so they think that stealing is no big deal and is justified?

**Rosa:** They should still know it's wrong by their inner conscience.

**Leif:** But it seems, doesn't it, that our consciences are shaped by our environment, by the people around us and the way we see those other people acting. Some children are born to parents who struggle with substance abuse, or who steal to get what they want, or who treat their own children very badly.

**Rosa:** Yes, I do see that we are influenced by our parents and by our environments. I don't deny that. But I think we can still rise above bad circumstances and make good or bad choices of our own free will.

**Leif:** You're saying you think that's true – that we make choices of our own free will – because people are morally responsible for what they do.

**Rosa:** That's right.

**Leif:** Aren't you putting the cart before the horse, in a way: you're assuming that people must have free will because you assume that they are morally responsible for (at least some of) what they do? That makes me curious. How do you support your belief that people are sometimes morally responsible for what they do?

**Rosa:** Well, I can see, just by looking around at our own society, as well as at societies across the world and through history, that people hold each other accountable. We obviously think that people are responsible to each other for their actions.

**Leif:** Okay, still, there are different theories about what the main purpose is of holding people responsible. I learned about this in my ethics class last year. There are competing theories about the central function of "moral responsibility ascriptions."

**Rosa:** Tell me more about that.

**Leif:** Our moral responsibility practices include praising and blaming, and rewarding and punishing, ourselves and other people. One idea about what we are doing, primarily, when we blame and punish people for their wrong actions is that we are attempting to steer their behavior in better directions in the future, and we are trying to encourage other people not to act in those ways.

**Rosa:** That makes sense.

**Leif:** That is a forward-looking view about the point of blaming and punishing people for their bad behavior. On that sort of view, the blaming and punishing practices of ours serve the useful purpose of getting people to behave better going forward in time. Think of the example of sometime when your parents took away a privilege of yours, maybe permission for you to use the family car to go out with your friends, after you did something that was against their rules. According to this forward-looking way of looking at what your parents were doing in blaming you and taking away your car use, your parents were trying to shape how you act in the future, to get you to stick to the family rules. They might also have been trying to set an example for your younger siblings, to mold their behavior, too. And in a case in which you did something good, like worked extra hard on an assignment for school, they might have praised you and rewarded you with a treat or privilege. On the forward-looking view of the *point* of praising and rewarding, your parents' main aim was to encourage you to keep up the good work and to keep being a good role model for your younger siblings.

**Rosa:** I can see that.

**Leif:** A different idea is more backward looking. On this viewpoint, what we are doing centrally when we praise or blame and reward or punish people is that we are trying to give them treatment that they *deserve* based on how they've already acted, whether they've acted rightly or wrongly. With respect to blaming and punishing, you could think of it like we are aiming to tip back the scales of justice. What I mean is that the wrongdoer has harmed a victim, and in response, we impose a penalty or harm on the wrongdoer, and we see the wrongdoer as deserving of the blame and punishment we dole out because of what he did wrong. On the flip side, when we praise and reward people for their good actions, according to this backward-looking viewpoint, our central purpose is not that we are trying to get them to do more good things in the future, but instead we are giving them some benefit they deserve for having benefitted the world (or society or someone in particular) by the good actions they did.

**Rosa:** I see. That makes sense. And there are other views, too?

**Leif:** Yes. Another interesting perspective is that when we praise and blame and reward and punish people, we are *expressing* something, such as an emotional reaction, like anger or resentment when someone has behaved badly, or admiration or gratitude when someone has done something kind and helpful. We might also be expressing a kind of statement about ourselves, like "*that* (bad action) is not something I do," or "I really want to be a person who does things like that (good action)."

**Rosa:** Okay, so let's talk about how these different views about our main purpose in holding people morally responsible – blaming and punishing people for wrong actions and praising and rewarding people for right actions – are connected to my statement that free will exists because we hold each other morally responsible.

**Leif:** Sure. Let's start with that view I just mentioned, which says we are expressing certain attitudes, such as resentment or gratitude, when we blame and praise people. It's called "an expressivist theory of moral responsibility," and it derives from the work of the philosopher Peter Strawson. It's quite normal that we experience these kinds of emotional attitudes in our interactions with people. Strawson called them "reactive attitudes" because they are attitudes we have in reaction to the attitudes that people display toward others or toward us in the ways they act. When their actions show that they respect and cherish us, we might feel love or gratitude, and when their actions show that they have disregard for us, we might feel hatred or resentment.

**Rosa:** So, to apply the theory to my example, in blaming my boyfriend for showing disrespect toward me when he (hypothetically) cheated on me, I'd be expressing my reaction of outrage or indignation.

**Leif:** Right. Now think about a case in which you do not take a reactive attitude toward someone after they've done something that caused harm – like when a very young toddler who is just starting to walk unsteadily knocks over a vase. Even if in a very fleeting way you feel anger, quickly that dissipates when you focus on who did it: a little toddler is not viewed as the appropriate subject for you to express anger toward by blaming her. Instead, you shift immediately into more of an "objective attitude," which is characterized not by reactive emotions but rather by your being in a calmer management mode regarding the situation – you remove the toddler so she doesn't get cut on the broken pieces, and you get some towels and bring a trash bag. Although having and expressing our reactive attitudes toward people in praising and blaming them is a big integral part of human interactions, the little toddler is exempt from these attitudes, since we see her as not the fit target of them.

**Rosa:** Okay.

**Leif:** In some other cases, too, you might take an objective, rather than reactive, attitude toward someone, such as when they are suffering from a mental illness that puts them out of touch with reality. In a different kind of case, you might see someone as the kind of person who ordinarily is the fit target of your reactive attitudes, but in some particular instance you don't blame him for what he did wrong because he has some excuse, such as that he didn't mean to do it or he didn't know better. But aside from these outlier cases in which someone is exempt or excused, we react to each other (and even to ourselves) nearly all the time with attitudes like pride, resentment, indignation, anger, contempt, love, and gratitude – and our practices of praising and blaming, and punishing and rewarding, give vent to these reactive attitudes.

**Rosa:** So, in most cases in human life, we hold each other responsible, like I said.

**Leif:** You also said that this shows that people have free will.

**Rosa:** Right, it does.

**Leif:** Well, then we need to talk more about what you mean by "free will." Here's a key point Peter Strawson makes: even if everything about our genetic makeup and our lives in the past leads up to exactly what we do when we do it and, when we act, we could not have done anything other than that action in exactly those circumstances, we are not by virtue of that fact thereby *exempt* or *excused*. We human beings still react to others and to ourselves, as usual, with the emotional attitudes we've named, and we give vent to or express these attitudes with praise and blame. This shows that moral responsibility does not really require "having options" in the way you have seemed to have in mind as we've been talking about free will.

**Rosa:** Hmm, I don't know about that. Honestly, I don't know if that expressive theory really captures the core of what I mean in saying that my boyfriend would be blameworthy if he were to cheat on me.

**Leif:** Well, then let's talk about that forward-looking theory of moral responsibility I mentioned. It would say that when you blame your boyfriend for cheating, you are trying to get him to be faithful to you in the future. For you to sensibly aim to achieve that purpose, it does not seem to be required that he had genuine options when he cheated – by "genuine options" I mean alternative actions he could have done at that time and in those circumstances, holding fixed the natural laws and everything that had happened in the past up to that point. It would still make good sense for you to blame him, on the forward-looking theory, even if he could not have resisted cheating at the time he did it, since your *goal* is to mold his future behavior by way of some negative experience you give him now – like yelling and withholding your affection – so that he will be more likely to stay faithful to you the next time an opportunity to cheat comes his way. You're attempting to shape the ways he acts going forward.

**Rosa:** I understand, but honestly I don't think that theory captures what I had in mind either. I mean to say that my boyfriend would *deserve* my blame and even some kind of punishment, like hearing my wrath, and not getting my kindness or attention for a while, or losing me altogether. I think I tend to believe in that backward-looking theory of the function of moral responsibility ascriptions you described. And I think his deserving blame and punishment requires that he acted with free will.

**Leif:** Okay, so let's say we agree to focusing on a backward-looking sense of moral responsibility, according to which praise and blame, and reward and punishment, are justified only if they are deserved on the basis of a person's actions. There are still varying accounts of what is required for being morally responsible in that "desert" sense (not "dessert" like cake, ha ha).

**Rosa:** I get it. I think you cannot deserve praise or blame, or reward or punishment, for what you do unless, when you choose to do it, there are options in front of you – multiple ways going into the future – and it was up to you which future to make the one that actually happens.

**Leif:** We learned in my class that that's called a "libertarian" view of free will, one that requires genuine options going into the future: multiple open pathways in front of the person as she chooses which action to do or which decision to make. It's not libertarian in the political sense, but in a metaphysical sense. We could talk more about metaphysical libertarian theories of free will another day. What I want to say now is that not all accounts of the free will required for being morally responsible for what we do in the sense of deserving praise or blame (or reward or punishment) are libertarian theories.

**Rosa:** So, you mean, some philosophers think that we can be deserving of praise and blame even if we do not have genuine options available to us for how to act – or forking pathways into the future – when we act?

**Leif:** That's right. Some theorists think that you can deserve blame when you did something wrong, and you can deserve praise when you did something right, if, when you did it, you could have done otherwise not in the sense that there were multiple forking pathways into the future in front of you at the time, holding fixed everything that had come before, but rather in a different sense. That different sense is this: to say that you could have done otherwise at the time is to say that you would have done otherwise if you had wanted to, or that you would have done otherwise if you had chosen to do otherwise.

**Rosa:** So let me see if I understand. Their idea is that the "ability to do otherwise" than act in the way someone does when she acts freely is something like the skill to do something different and the dependence of what she does on her desires or her will. If she had wanted to do something else, then she would have, and that captures the idea that she could have done otherwise.

**Leif:** I think that's about right. It's called a "conditional analysis of the ability to do otherwise" because it defines the relevant ability to do otherwise in terms of a conditional statement: *if* she had wanted or chosen to do otherwise, then she would have. The idea is associated with the philosophers Thomas Hobbes and David Hume, and it has more contemporary proponents, too.

**Rosa:** I see. So those philosophers don't think that deserving blame or praise requires having alternative forking pathways into the future available to you at the time you act, keeping fixed all the natural laws and events of the past.

**Leif:** Right. Here is another important theory of what it takes to have the kind of control over our actions that we need to be morally responsible for them (in the sense of deserving praise and blame), developed in the work of the philosopher John Martin Fischer. It is not a libertarian theory. Think about the "mechanisms" that produce our actions. For example, have you ever seen one of those demonstrations on a stage where a hypnotist puts someone into kind of a trance and plants a trigger word in the mind of the subject?

**Rosa:** Yes, I saw one of those once.

**Leif:** Suppose that, while the subject is under hypnosis, the hypnotist tells him that when he hears the word "elephant," he is going to raise his hands in the air, wave them around, and start making blaring noises like an elephant. And then the hypnotist brings the subject out of the trance, and the subject goes back into the audience to his seat. The hypnotist keeps talking, then suddenly inserts the word "elephant" into his sentence and, sure enough, the subject in the audience starts waving his arms in the air and making the ridiculous elephant-like sounds, and the rest of the audience bursts out laughing.

**Rosa:** I'd hate to be that subject.

**Leif:** Yeah, it would be mortifying! Anyway, in that case, the mechanism that produced the subject's hand waving and blaring noises was post-hypnotic suggestion, and it was an action-producing mechanism that was completely insensitive to any reasons the subject might have for doing that or not doing that. Presumably, the subject would not want to embarrass his friends and would not want to draw attention to himself and would think it is rude to interrupt a speaker on stage. Those are actual reasons he has for watching attentively and for not making a scene by ridiculously blaring like an elephant. But the mechanism that led to his behaving that way was not at all attuned to those reasons of his – it did not take them into account or respond to them or react to them at all. It was not a "reasons-responsive mechanism."

**Rosa:** Okay, right.

**Leif:** Another example is this: imagine a puppeteer controlling the limbs of a marionette. The mechanism that produces the movement of the marionette's arms and legs is the contraption pulling on the strings by the wooden handle controlled by the puppeteer. That is not a reasons-responsive mechanism leading to movement.

Neither is the force of a true addiction, when someone with a severe substance use disorder keeps on using a drug, and would keep using it and not cut himself off, no matter what reasons he might consider for stopping.

**Rosa:** Got it.

**Leif:** By contrast to those non-reasons-responsive mechanisms, think about when you reflect on what major to declare in college, mulling over the reasons favoring art and the reasons favoring finance. Suppose that when you decide to declare art as your major, your reflection was your own in a way that took into consideration your various reasons, including your passions and values, and you would have chosen finance had there been good enough reasons from your perspective favoring that, so that the mechanism that led to your action of declaring the art major was responsive to reasons. You acted with the kind of control needed for you to be praiseworthy for doing that (supposing it is morally good that you chose the art major, although it's probably just morally neutral).

**Rosa:** That makes sense.

**Leif:** But, see, Fischer's account of this, which he calls having "guidance control" over your action, does not require your having forking pathways into the future available to you. What he calls "regulative control" requires being able to do otherwise in the exact circumstances – forking pathways into the future – but regulative control is not needed for moral responsibility, Fischer argues. Guidance control is what is required for moral responsibility.

**Rosa:** Hmm. I'll think about that.

**Leif:** Here is a different account that is not libertarian and that does not require genuinely available alternative options for all free actions for which you can be morally responsible. On the view of the philosopher Susan Wolf, to act freely in the sense required for moral responsibility is to have the ability to act in accordance with the True and the Good. So, when you do something wrong, in order to deserve blame for that, you do have to have been able at the time to have done something right, since you have to have been able to act in accord with the True and the Good. Blameworthiness for doing what's wrong requires being able to do what's right. But when you do something that's morally right, you acted freely in doing it if you acted from an appreciation of the True and the Good – so you are praiseworthy for it, even if you could not have done something wrong at that time, instead. This view has a kind of "asymmetry" involved in it regarding the requirement of available options. The philosopher Dana Kay Nelkin has a similar view, which says that the freedom needed for moral responsibility is the ability to act rightly from an appreciation of right reasons.

**Rosa:** It sounds like it was an interesting ethics class. I should take it sometime.

**Leif:** Yes, the professor was great, too!

**Rosa:** To recap where we are, my point was that we have free will because we are morally responsible for a lot of what we do, and being morally responsible for what we do requires that we do it freely. Now you've said that holding each other morally responsible for what we do can be made sense of in different ways – as expressing our natural reactive attitudes, or as aiming to shape future behavior, or as aiming to give people the treatment they deserve based on what they've done that is either right or wrong. And you've said that, on a conception of moral responsibility rooted in the idea of what people deserve due to what they have freely done, there are different accounts of what the "freedom" is that people need to have for it.

**Leif:** Exactly.

**Rosa:** I can still say that we hold each other morally responsible in the deserving sense, and that free will is required for that, and so we have free will.

**Leif:** Yes, you can say that, of course. It seems like we have left open, though, what sense of free will it is that you are arguing that we have. The varying accounts of the kind of freedom required for moral responsibility require that different things would have to be true about us and true about the world, for us to actually have that freedom. For instance, if John Martin Fischer is right, then we must be able to act sometimes from reasons-responsive mechanisms in order to be morally responsible for (at least some of) what we do. And if Susan Wolf is right, then we must be able sometimes to act in accord with the True and the Good, which requires that we sometimes understand what is true relating to our situations and what it would be good to do in those situations. And if libertarians are right, then the universe must allow for there to be forking pathways into the future in front of us sometimes, holding fixed the past and natural laws, and it has to be sometimes up to us which one of those possible futures becomes the actual future.

**Rosa:** I am still convinced that we are sometimes morally responsible for what we do and so that we are free in at least one of those plausible senses of what that means.

**Leif:** Let's go for now, Rosa, if that's okay with you…I need to eat dinner and want to mull over all this some more. See you soon?

**Rosa:** Sounds great, how about in a couple days – let's get in touch about when would be a good time. Have a good night, Leif.

# Day 3: Does Our Humanity Show that We Have Free Will?

Abstract

*Today Rosa takes a different tack, arguing that our humanity shows that we have free will. She states that the power of free will is a distinctive feature of human beings, one that differentiates us from other animals. As they explore this idea, Leif is not impressed, so Rosa changes direction somewhat. She next suggests that it is not species membership that shows that we have free will, but rather the fact that human life has meaning. If we did not have free will, then nothing we do would matter and there would be no purpose to our lives, Rosa reasons, so we must have free will. Together Leif and Rosa explore her ideas about motivation, fate, and passivity, allowing them to distinguish between agency and free agency. They go on to explore different accounts of meaning in human life and the requirement, or lack of it, of free will for meaningfulness.*

**Rosa:** I think human beings have free will because we are different from other animals.

**Leif:** Could you say what you mean by that – how are we different?

**Rosa:** We are intelligent, and we can reason and make free choices to help or to hurt each other, based on reasoning and our values.

**Leif:** I see. But don't some other species show evidence of making choices to help or hurt each other? Elephants show empathy. Dogs can choose to run and bark to

DOI: 10.4324/9781003684268-5

get help when their owners are hurt. There are some amazing stories about how pets have helped people and saved their owners' lives.

**Rosa:** That's true, but our human choices are more complicated, like which college to go to or which job to take.

**Leif:** That wouldn't really show that we have free will, would it? It seems to me that would only indicate that we are complex and intelligent beings who have constructed schools and businesses. Yes, I agree, we can do different things than some other species, but I don't see how that shows that we have free will. Some members of other species can do things we can't do, like fly on their own power (without a plane), as owls can, or run fifty miles per hour, like jaguars can.

**Rosa:** I'm not saying we are better than all other species in every way, or that we have more of every kind of power than they have. I realize we can't run as fast as jaguars! What I'm saying is that we do have this special power of free will as human beings, a power that members of other species don't have.

**Leif:** I think I'm not getting the reason you think we have that power of free will.

**Rosa:** I'm trying not to say what I actually think to back this up, which is that God made us human beings in His image, and one thing that means is that, in addition to making us rational and spiritual beings, God gave us free will. I know we said we would set aside my religious beliefs as reasons. Maybe I can just put it this way, without referring to God: I think human beings are especially valuable, more valuable than, for instance, beetles and frogs, and one reason for that is that we are living beings with a power that is unique and special, a power for making free decisions of our own. We can decide how to direct our lives, rather than acting merely on instinct.

**Leif:** Maybe I see what you are getting at: lions hunt prey by instinct, for example. But we human beings do not have to act only by our natural instincts. Maybe it is an evolved instinct of ours to eat meat. But we human beings are special among living things in that we can freely decide to become vegans, for instance. We don't have to follow our instinctual drives.

**Rosa:** Yes, exactly, that's what I'm pointing to, an ability we human beings have, which it seems that other kinds of animals do not have, and one that makes us especially valuable as humans.

**Leif:** I guess I'd say, though, that not everyone thinks we are more valuable than other living beings – some people think that all biological life has value and deserves respect and care. Some people think that cows are very valuable, even sacred.

**Rosa:** Well, I just don't agree with them. I think it's perfectly fine to kill termites and mosquitos, for instance, and obviously it's fine to kill bacteria and viruses that infect us – and, by the way, I think it's fine to eat meat, especially pasture-raised. Not all biological life has value, but human beings definitely do have value.

**Leif:** What do you think makes it true that we really *are* more valuable, rather than our just thinking we are (when we are not) because that's what we want to believe, and it makes us feel important to think that?

**Rosa:** You know, once again, I'm having to resist appealing to my religious beliefs. I think God made us human beings above the other animals and told us to rule over the created world. And God lets us freely choose to obey Him or not, which explains why we are especially valuable. God gave us a power that He did not give to other species – he made us higher than them and more like Himself.

**Leif:** Do you think that your idea can be backed up without appealing to religious scriptures or traditional religious doctrines?

**Rosa:** Let me see…a few minutes ago I said that we human beings can make choices based not just on our natural instincts but based on our reasoning and values, and you said that you see the point that some animals in the wild hunt prey by instinct – lions, for example – but we, by contrast, can go against our instincts by deciding to become vegans. That is the observation I am pointing to as evidence that we human beings have free will.

**Leif:** So, you have in mind that we can make judgments about what it would be good for us to do or what would be morally right for us to do in our circumstances. And we can act from our values or from our sense of what our moral duties are, rather than always acting in every instance from our biological drives and impulses. Going back to your example the other day about your boyfriend, because he is human, he can choose to keep his obligation to stay faithful to you as he promised, rather than acting on a physical impulse to do something that would hurt you and violate his promise.

**Rosa:** That's right.

**Leif:** To be honest, I'm still not seeing how that is an argument that we have free will because we are human. It seems like you are describing what you think free will is, and it involves being able to act from reasons that we understand to be right reasons for doing something. It seems like you're characterizing free will, rather than saying *why* we should think that we do have it.

**Rosa:** Well, I think all human beings have that ability you just named, to act from reasons that they understand to be right reasons.

**Leif:** I'm kind of scratching my head wondering, how is that not just an assertion that we have free will? We are looking for a case for thinking that we have it. I'm hoping to hear evidence or an argument.

**Rosa:** Yes, I know, I'm trying. What about this: we human beings are creative. We have created great works of art, like Michelangelo's frescos on the ceiling of the Sistine Chapel, and the gorgeous paintings by Monet, Gauguin, and so many other artists, both famous and amateur like me (not saying my own paintings are gorgeous, but I try!). Human beings create stunning sculptures and achievements in architecture. We make creative breakthroughs in science, and we can write poetry and fiction, compose music, and create films and theatrical productions. We are amazingly creative, and I think this shows that we human beings have free will.

**Leif:** Hmm. Why would someone have to have free will in order to be creative?

**Rosa:** Our creativity, enabled by our free will, makes us different from machines and robots. As human beings, we can use our free will to come up with new inventions, to think of new ideas, to choose to branch out in new directions that haven't been thought of before. We don't just carry out programing that has been put into us or just channel forces from beyond or outside us. We creatively chart our own courses through life. Our human creative achievements show that we have free will.

**Leif:** I agree with you that we human beings are creative, Rosa. I suppose whether or not that requires free will depends on what you mean by "free will." The great painters and composers of the world may have been channeling forces from beyond them in some sense, such as in the sense of being inspired by the artistic works of others, or in being driven by a passion to realize their artistic visions despite their physical fatigue, or in the sense of being animated by a muse or motivated by love. I don't think any of that being true would rule out their artistic works counting as instances of true creativity. We human beings can create paintings, sculptures, doodles, mathematical proofs, and new recipes, I agree, and all of that seems to hold true even if there are causes ultimately outside of us, like what we've read and seen in the past and how we were raised, that led to our doing what we did.

**Rosa:** From my perspective, to count as acts of genuine creativity, our actions have to come from us and not from outside us. They need to be original, which indicates free will.

**Leif:** When you make a painting, say of the scene in front of us here, do you think there would have to be no causes of what colors you choose to use and where you choose to put them on a canvas?

**Rosa:** Well, there could be reasons for why I pick the specific colors and reasons for why I put them where I do, but I think my painting being creative requires that it comes from my genuine free choices, meaning that I could have picked different colors and could have put them in different places than I did.

**Leif:** It seems to me that, when you create a painting, all sorts of experiences lead up to you painting it in the way you do, like lessons in your art classes and your memories. Your artistic vision and skill have been shaped by your instructors and by the works of other painters you've seen. The way I see it, creativity does not by its very nature require bringing something out of nothing, and it doesn't require you being the sole uncaused cause of what you make. Instead, creativity involves putting together thoughts or colors or ingredients or pieces of metal in what are, to the person who is creating, new ways, in an endeavor of discovery and making something. We can put together thoughts and ingredients and colors in ways we ourselves have not done before and maybe in ways that others have not done before if our creative works show advancement in our creative field. But it's not clear to me what kind of, if any, free will this involves. At any rate, you started out today arguing that our humanity shows that we have free will, and I'm not sure that creativity is unique to humanity, anyway.

**Rosa:** Why do you say that?

**Leif:** I keep bringing up other species, but don't you think that members of other species show creativity in figuring out how to solve problems, like where to hide to avoid predators, and for some species, even how to open doors? I don't know if there are intelligent beings on any other planets in the universe, but for all we know, if there are, they could be creative, too. And I'm not sure that machines couldn't become creative, if they aren't already. It seems like some complex AI systems already are creative!

**Rosa:** What is your point, then, Leif?

**Leif:** I don't think it works to say this: we are members of the species *homo sapiens*, and therefore we have free will. It seems like our species membership is defined by our DNA, by our shared genetic code and our ability to interbreed and produce viable and fertile offspring, as biologists say. And our being members of one species rather than another doesn't really tell us whether or not we have a power to make free choices.

**Rosa:** Well, what about simply the fact that we are especially valuable as human beings, which is explained at least partly by our having free will?

**Leif:** I don't think it is obvious that we are especially valuable, as opposed to just wanting to think that we are. Even if beings that can act with empathy and care for others, or beings that are rational or intelligent, have more value than beings that cannot or are not, I'm not sure that breaks down along the lines of species membership and includes only human beings. I mentioned dogs earlier, who can go get help for their owners rather than ignoring their owners' distress, and elephants who show empathy and mutual care. Chimpanzees and pigs and dolphins show intelligence, too. I just don't see it as a persuasive reason for concluding that we have free will to say, well, we are human. It seems like lurking in the background is your idea that God made human beings special and gave them free will.

**Rosa:** I guess I see what you mean. Okay, let me try a little bit different tack.

**Leif:** Sure.

**Rosa:** How about this argument, Leif. I think human life has meaning, and if we didn't have free will, then nothing we do would matter and there'd be no purpose to our lives. So, we must have free will.

**Leif:** Wait, let me see if I understand what you're saying. You think that we need to believe that we have free will for us to have a sense of purpose in our lives. Could you say more about that?

**Rosa:** Yes. If I did not think I had free will, then I'd lose all motivation to do anything, because whatever was going to happen would happen no matter what I do.

**Leif:** Without a belief in free will, you wouldn't be motivated to do anything? You'd have an attitude of resignation toward life, and you'd feel helpless or passive?

**Rosa:** Yes, something like that. My belief that I have free will helps get me going every day, to make choices about how to spend my time and choices about how to direct my life. If I didn't think that the direction of my life is up to my own free choices, then I would just give up and wait for destiny or fate to unfold.

**Leif:** That's confusing to me. Are you saying that if you didn't have free will, and you were to believe that you didn't have free will, then you'd just stay in bed all day and not do anything?

**Rosa:** Well…yes! Because whatever is going to happen would not depend on my free choices, so I'd just wait for everything to happen instead of making anything happen by what I decide to do.

**Leif:** If you stayed in bed tomorrow morning because you gave up your belief that you have free will, wouldn't that be a choice you made, to stay in bed, instead of going out to do whatever you had previously planned?

**Rosa:** Yes, I guess it would be a choice to stay in bed, but if we don't have free will then it wouldn't be a free choice. It would just be what I did.

**Leif:** You think, then, that we can make choices even if we do not have free will, or even if we believe that we do not have free will.

**Rosa:** I guess in a sense, yes, we would still "decide" what to do, without free will – like I could make up my mind that I'm going to just stay in bed – but we wouldn't really be deciding *among options* we had available to us, and it wouldn't matter what we decided.

**Leif:** Let's talk first about that idea that "it wouldn't matter" and then about "deciding among options."

**Rosa:** Okay. By saying "it wouldn't matter," I mean that, if we don't have free will, then some grand plan of the universe just unfolds, like fate or destiny – or divine providence, but I'm trying to leave aside religion. If I do not have free will, then I can't really make any difference to a grand plan or fate or destiny. Things are just going to be what they're going to be, and none of it is really up to me. So, nothing I do would matter.

**Leif:** Wait a minute. If you stay in bed every day all day, then you're never going to become a great artist or even make a living as a working artist, as you've told me you want to do. If you were to give up on art and stay in bed every day, then being an artist could not be your fate or destiny.

**Rosa:** Well, then my destiny would be that I am a person who stays in bed all the time.

**Leif:** But I don't see why you think we would need to be passive all the time, never doing anything we want to do or pursuing any goals, if we didn't have free will and if you didn't also believe that we have free will. Maybe this would help to explain what I'm getting at. Think about the difference between an "agent," on the one hand, and a "patient" or "object," on the other. You can be an agent – someone who acts – as opposed to a patient or an object – something that is only acted upon or pushed about by other forces, like a falling leaf in the wind. Your being an agent and exercising your agency in acting does not require that you ever act freely. Falling leaves and rocks can't act at all, but you can act, even if none of your acts count as free acts.

**Rosa:** Okay, yes, I guess I see what you mean: I am more than an object, and I am not just acted upon. I can do things by will.

**Leif:** Right, and we are trying to figure out if it is reasonable to think that you and I ever do anything of our own *free will.*

**Rosa:** Well, I was saying that, if we don't ever do anything of our own free will, then fate or destiny is just going to unfold, no matter what, and that would undermine my motivation in life and my sense of purpose – my sense that my life matters and has meaning.

**Leif:** As far as "fate" and "destiny" go, I think it's hard to figure out what those words mean. But I do think that the way your life develops depends partly on your actions. I don't think it really makes sense to think that there's some endpoint to your life (your "destiny") where you're going to end up, no matter what you do. And don't you think that your life matters – matters more than the life of a flea, for instance – regardless of whether or not you have free will? It matters to your parents, and to your brother, and to your boyfriend, and to everyone who cares about you.

**Rosa:** Well, yes, I can see that my life matters to those people, just like your life matters to the people who care about you. But I still don't think that human life as a whole has meaning if we don't have free will. And since I think that human life does have meaning, I think that we have free will.

**Leif:** Could you say more about that?

**Rosa:** Human life as a whole – humanity's existence – has meaning only if someone put us here, someone like a divine being, and only if we were put here so that we could freely decide whether or not to carry out that purpose we were put here for.

**Leif:** Your religious beliefs do seem to permeate your thinking.

**Rosa:** Yes, they do…but isn't that to be expected? Believing in God has huge implications. It affects what I believe about morality, and what I believe about the value of human beings, and what I believe about the beginning of the universe, and what I believe about why things happen in the ways they do happen on Earth and in the afterlife. It's really hard to just completely set aside my religious beliefs in these conversations we are having about a philosophical topic that is so connected to all of these other things.

**Leif:** I can see that. It's probably that way for a lot of people, too. Still, what you said seems to suggest that human life overall has a purpose only if we were put here intentionally for some reason by a divine being. But that is different from the claim that the existence of a meaning of human life overall implies that we actually do have free will.

**Rosa:** I can see your point that whether or not someone thinks there is a meaning of human life overall, a purpose for which we exist, can depend on whether or not they think that God exists, maybe without it depending on whether or not they think human free will exists. For there to be a purpose for which we human beings were created requires a Creator who made us for a purpose. But maybe there being a purpose to human existence in that sense does not require our having free will – although I guess it depends on the particular religious believer and what they think the purpose is for which God created us.

**Leif:** What do you mean by that?

**Rosa:** Well, some religious believers might think that God created us so that God could enjoy our company. That's why we human beings are here, for God to enjoy our presence in the universe. I can see that our having that purpose does not seem to require that we have free will, because God could enjoy us just like God enjoys the beauty and company of doves and butterflies. But different religious believers might think that the purpose of human existence is for us to freely choose to love God. In that case, they would think, like I do, that the meaning of human life requires that we have free will.

**Leif:** I see. So, in that case, the reason for believing in free will is really that love requires it, and then there's the added belief that the purpose for which God made us is for us to love God.

**Rosa:** I think that's right.

**Leif:** So maybe we could talk about the connection between love and free will next time we get together.

**Rosa:** Sure.

**Leif:** I wonder if we could stay on the topic of meaning a little longer today. I've heard a distinction made between the meaning *of* life – as in the purpose of human existence as a whole – and meaning *in* life.

**Rosa:** What do you mean by meaning in life?

**Leif:** The idea is that, even if there is no supernatural being who put us human beings here for some purpose the divine being had in mind, it does not follow that all human lives are meaningless. Some human lives seem to be more meaningful than others, and maybe we can capture that intuition we have about the difference between some lives and others by talking about meaningfulness *in* individual lives.

**Rosa:** Could you give some examples?

**Leif:** Yes – the lives of Nelson Mandela and Michelle Obama and William Shakespeare and Rosa Parks and Claude Monet seem to have, or to have had, a lot of meaning in them, wouldn't you say, as compared to the life of someone who stays at home alone playing solitaire on a computer all day, eating junk foods, never making any impact on the world.

**Rosa:** I suppose that depends on how you fill in the specific details of the case of the person who plays games on the computer. Maybe that's the only thing he can do in his circumstances, and he does it to keep up his mental health, because he is in constant pain from a disabling injury. Surviving his pain through distraction, getting through each day while being kind to his family members, could make his life meaningful. I don't see what healthy food would have to do with it, if sweets and salty snacks make his pain easier to bear. We shouldn't say his life is unimportant and lacks meaning just because he doesn't lead a great social movement or create lasting artistic masterpieces, especially when he simply can't.

**Leif:** You're right, I see your point there. I could fill in the case a different way. Suppose, instead, that the person we have in mind is a medical researcher who has spent decades working with a research team on creating new technology for developing vaccines. He is not injured, and he isn't in pain. He just leaves work one day, goes home, and for the next ten years, he spends every day at home playing solitaire on the computer, just to pass the time. He enjoys it somewhat, but not a lot, and he doesn't think the games he plays really matter. He doesn't interact with people, and he has no plans to return to his medical research. It seems to me that, now, the way he is living lacks meaning compared to the way he was living before.

**Rosa:** I guess I do share that perspective.

**Leif:** And then suppose he wakes up one morning and thinks, "Good grief, I've been wasting my life these past ten years! The way I am living now lacks meaning." That seems like a phenomenon that happens to people.

**Rosa:** Like in a mid-life crisis, when someone thinks, what have I been doing all this time? None of it means anything. Then sometimes people want to make changes and build a more meaningful life going forward.

**Leif:** Exactly. That's what I mean in talking about meaning *in* life. Even if there is no God, even if there is no overarching purpose of a divine being for which we human beings were put here, still, our lives can have more meaning or less meaning in them. Some people think of meaning in life as a purely subjective matter, so your life has meaning, for instance, just in case you're happy, no matter what you're doing or not doing. Other people think that meaning in life is determined purely objectively, so that, no matter how you feel about it or whether or not you enjoy it, if you are doing something that has a positive impact on the world – like working toward a cure for cancer or leading a social justice movement – then your life is meaningful. Other people have a hybrid viewpoint that involves both subjective and objective elements in the account of meaning in life.

**Rosa:** What's an example of a hybrid view?

**Leif:** A philosopher I mentioned last time, Susan Wolf, suggests that people's lives have meaning in them when they are actively engaged in projects that are positively valuable. Another way she expresses the view is that meaning in life comes from loving things (projects, people, activities, and objects) that are worth loving. There's the subjective aspect of being actively engaged in something, or loving something, and then there's the objective component of the theory, which is that *what* you're actively engaged in doing, or what you love, must actually have positive value.

**Rosa:** Why does she include the objective component?

**Leif:** It's because she sees lives that are actively engaged in projects that don't really matter, ones that do not have positive value, as lacking in meaning, such as someone who shops all day long, staying active but just to stay busy passing the time, or someone who actively lines up rocks day after day, not making any art or building anything, just lining them up over and over.

**Rosa:** I see, that makes sense. Well, let's get back to free will. I was saying that I believe we have free will because "human life has meaning." Suppose we take that statement to mean, human lives can have meaning in them – we can live meaningful lives, along the lines that Susan Wolf depicts them.

**Leif:** That helps clarify our topic of meaningfulness. The thing is, I don't think that Susan Wolf's theory of meaning in life requires that we have free will in the sense you have seemed to have in mind during our conversations so far.

**Rosa:** I have meant, I think, that we can sometimes make choices among options that are available to us as forking pathways going into the future, according to our deliberations over our reasons, including our beliefs and desires and values. Meaning in life requires that, don't you think?

**Leif:** If we go with Wolf's theory of meaning in life, then we do need to be able to love – or, put differently, to actively engage ourselves with – people or projects. And we do need to be able to tell what is worth loving – or, put differently, to discern what does have positive value. But your being able to actively engage in some project, like raising money for cancer research, does not require that you could have chosen to do something else instead in your very same circumstances, and it doesn't require that you could have done something wrong. And your being able to tell what is good and what is true, such as that it is positively valuable to raise money for cancer research, does not require your having an ability to choose among options that are physically available to you or the ability to choose to do horrible things, either. On Wolf's own theory of free action, your acting freely requires that you have the ability to act in accordance with the True and the Good. It does not require that you could have acted in a different way at the time you acted in a right way, and it does not require that you could have acted badly, instead.

**Rosa:** I see. So she thinks that meaning in life does not require what we've talked about as "genuine options."

**Leif:** Right. Here's another philosopher's view about our freedom in relation to meaning in life, which is relevant to this question about having alternative options. John Martin Fischer argues that, when a person acts freely, she writes a sentence in the story of her life, which gives her life the shape of a narrative and thereby provides her life with a unique sort of meaning and value. Meaning in life involves making a statement, Fischer suggests, rather than making a difference. Making a statement can be achieved by a person who acts with guidance control, on a reasons-responsive mechanism, which is compatible with not being able to do otherwise than act as one does in the sense of having multiple forking pathways available to one going into the future.

**Rosa:** Hmm, I'll have to think more about that. But I did notice that Wolf referred to love in her account of meaningfulness in life, and I think love itself requires the kind of free will I have in mind.

**Leif:** We'll talk about that next time we get together?

**Rosa:** Sounds great. Take care, Leif!

# Day 4: Does Love Show that We Have Free Will?

Abstract

*In the fourth day's conversation, Rosa suggests that we have free will because love requires it, and we can love each other. Because by now she has gotten used to Leif's objections, Rosa says right away that he cannot question this basis for believing in free will – as he questioned her religious grounds for believing in it – because, when she says she believes in free will since love requires free will and we human beings experience love, she is not assuming that God exists. Leif points out that Rosa is, though, asserting two things: one, that love exists, and two, that love requires free will. Rosa agrees, but believes those claims are obvious to virtually everyone, or at least they are not as controversial or difficult to back up as the assumption that God exists. Rosa and Leif explore the nature of love, various types of love, and the relations between love and free will.*

**Rosa:** Hi Leif! Let's get right back into it and pick up where we ended last time. Here's what I want to focus on today: I think we have free will because love requires it, and we can love each other. And let me say – because I've gotten used to your objections by now – I think you can't question love as a basis for believing in free will as much as you questioned my religious grounds for believing in it. That is because when I say that I believe in free

DOI: 10.4324/9781003684268-6

will since love requires free will, and we human beings experience love, I am not assuming that God exists.

**Leif:** That's true. You are, though, right, assuming, or at least asserting, two things: one, that love exists, and two, that love requires free will.

**Rosa:** Okay, yes, but I think those are obvious to virtually everyone. I don't think those ideas are nearly as controversial or difficult to back up as the assumption that God exists. Let's start with the first one, that there really is love. You don't doubt that, do you?

**Leif:** Well, that depends on what you mean by "love."

**Rosa:** One thing I mean is that obviously we fall in love – sometimes with a person, and sometimes with a place we've discovered, and sometimes with some new sport or hobby, like when you take up watercolor painting and find out that you just love it.

**Leif:** It's interesting that you use those examples of falling in love, because those seem to me like instances of passion, as in romantic love and in experiences when someone becomes completely taken by a new interest. Doesn't the language of "falling" suggest that it's something that happens to you, rather than something you freely choose? It's not like you freely decide to find yourself suddenly loving painting, or soccer, or some new place, or some object, like your baseball collection or your new pair of shoes. It's more like you discover yourself feeling really wonderful about them, feeling attached to them and passionate about them – it's not a matter of making a free choice to feel passionately.

**Rosa:** I can see that the language of "falling in love" sounds like falling off a cliff or getting overcome by passion, and I guess it's true that falling in love, especially romantic love, can feel like something that happens to you. It's like getting caught in a spring rain or like getting bathed in warm sunshine when you walk out of a cold building. But obviously not all love is a feeling that just happens to us, and that feeling is not even all of what makes up romantic love.

**Leif:** What else do you think is involved in romantic love? And what other kinds of love do you have in mind?

**Rosa:** In romantic love, if it's going to work, we each need to make decisions, for instance, choices about the ways we connect and how often to communicate, and choices about how to spend our time together. In romantic love, we also make free choices about whether or not to form commitments to each other and what kind of commitments. We freely choose what kinds of promises to make and what kinds of expectations we want to share concerning each other. Those choices and those commitments we make involve our free will. They don't just happen to us. We weigh up lots of factors, and we make an effort to decide about those kinds of relational matters.

**Leif:** Okay, but I'm not sure we know that those choices and decisions you are talking about are free ones. I think there are lots of causes that explain why we fall in love with the people we fall in love with, and there are causal explanations for why we make, or don't make, the commitments we make to other people. It seems to me a lot of it can be explained by how we were raised, and by the kinds of relationships we have seen modeled by our parents and our relatives and in our broader culture, such as in films and literature. Maybe not just a lot of it can be explained, but all of it. Maybe the commitments we happen to make to particular people (or our lack of commitments) is inevitable, given everything about us and our past experiences.

**Rosa:** I'm not sure what to say about that. But let me answer your question about the other kinds of love I have in mind. There isn't just romantic love. There's also the love we have for our friends. And there's a kind of general love for fellow human beings that a lot of us experience and try to cultivate in ourselves. Also, there is the love that good parents have for their children, and the love that children have for their parents. There's sibling love, and love for our ancestors and descendants – grandparents, aunts, uncles, nieces, nephews, cousins, and others. You get the point: there is familial love. There's so much more. There is love people have for sports teams and love for being in nature. There are many kinds of love, and there are different aspects of love.

**Leif:** So, you think some loves involve passion, and some involve commitment, as you said regarding romantic love. What is love like in these other kinds of love you have in mind?

**Rosa:** I think in some cases it involves admiration, and it should involve respect. There are feelings that are part of it, and those feelings are different depending on who or what is loved, such as if it's your brother or your infant or your favorite athlete (who probably doesn't even know you). I guess the last one is kind of like a crush – being a fan involves admiration, and it also includes a kind of commitment to caring about how they do, hoping that they win in competitions, and watching them play.

**Leif:** To pick up on the feelings point some more – the loves you've mention do seem to involve certain feelings: in some cases, feelings of excitement (like in romantic love), in other cases, feelings of warmth and pleasure (like when you're with a close friend), in other cases feelings of admiration (like for your parents, or for your heroes), in others feelings of pride (like in your children or in your siblings), and in others feelings of comfort (like when you're with a group of people who accept and understand you). All those feelings seem to be things we experience and, again, seem to be things that happen to us, like feeling too cold or too hot when we're trying to fall asleep, or feeling awe at seeing a beautiful sunset.

**Rosa:** I do think love is more than a feeling, and I think that we can freely choose to do things that play a role in what feelings we have. For example, if I freely choose to buy my friend a gift when I see something I think she might like, and then I freely give it to her rather than keeping it for myself, I get feelings of pleasure and satisfaction, and my feelings of affection for her and my attachment to her get stronger. Those feelings didn't just happen to me but instead they came from free choices of mine.

**Leif:** I can see that. What we do has effects on how we feel. When I drink too much coffee, I feel jittery. When I stay up too late, I feel unmotivated the next day. But we have not established yet that anything we do is done by our own free will. Take the case of when you buy your friend a gift, in your example. We, you and I in this conversation here, have not yet found a reason, or anyway a strong case, for thinking that you buy a gift (or do anything else) of your own free will. Today you are suggesting a new reason to believe that we really do have free will. The reason is that love is real, and it requires free will. So to move forward with trying to think through that suggestion, let's talk some more about love itself. Can you say in a nutshell what you think love is?

**Rosa:** Well, I think it typically involves some feelings like the ones we've mentioned, feelings of affection, closeness, pleasure in their company, admiration for them, satisfaction at being accepted by them. Love also involves some desires.

**Leif:** What ones?

**Rosa:** When you love someone, you have desires to spend time with them and to be close to them. And you want the person you love or the object of your love to do well, to flourish. Part of my loving my plant collection, for example, is wanting my plants to stay green, appropriately watered, and vibrant. I want them to do well as plants. Part of my love for my boyfriend is wanting him to have the best life that's possible for him – for him to stay physically well and to be emotionally healthy and to flourish as a person in every way, in his career, in his family life, and in his projects.

**Leif:** So, there are feelings and desires that you think of as typical aspects of love. Anything else that helps describe love?

**Rosa:** Actions, too – we wouldn't believe someone if they said they love another person if they never do anything to benefit that person at all. If they never help that person or listen to that person or show that person affection or care, we think they don't really love them. Loving parents take care of their children – they buy them food and cook them meals and provide them with a place to live, if they can, and they act in ways that show care for their children's emotional and intellectual well-being, too. Loving parents don't just feel warm feelings toward their kids, and they don't just want them to be well. They also do things to express those feelings and to help their children to live as well as possible.

**Leif:** I would agree to all that. And earlier you mentioned commitments – people who love another person, or a cause, or a pet, often make commitments to who or what they love, commitments to do certain things.

**Rosa:** Right.

**Leif:** So when you say that love obviously exists, you mean to say that some relationships and experiences you have yourself – as well as experiences and relationships that you observe other people having – have the features you've been elaborating. They count as relationships of love and experiences of love because they have the characteristics that are definitive of love.

**Rosa:** That's right. I guess another way to say it is that not all the interactions I see human beings having, including myself, are interactions of hatred or indifference. Some of them are loving. That seems to be a fact about the world. Many parents love their children. Many people love their partners. Many people love their pets. Many people love their friends. That does not seem to be controversial or unclear, the way it is controversial and unclear whether or not God exists.

**Leif:** Okay, so if I asked, are you sure those are *really* instances of love, not just people who are acting as if they love the other person – or if I put the question as, does *real love* exist – you'd say it's a silly question.

**Rosa:** Yes, because there's a real difference between relationships that are unloving, on the one hand, and relationships that are loving, on the other. It might be hard to assess whether or not a certain relationship is a loving one, sometimes because we don't have enough information, and also because it's hard to define love precisely. There might be a cluster of cases that all count as love without them all sharing exactly the same features. But there are definitely real cases of love, including the real love I have for my brother. Nobody could convince me that that's not a case of real love. I grew up with my brother, sharing tons of experiences, and I'm so glad I have him in my life. I cherish him. I want everything to always go well for him. I want us to be close and to share special occasions and events in our lives for as long as we're alive. There's no way that love isn't real.

**Leif:** Let's say we both agree to that. Love exists. There are relationships that really count as relationships of love. Now on to the second key part of your case: why do you think love requires free will?

**Rosa:** Because it's not real love if you *have to* love someone else, if you have no choice about it.

**Leif:** So, what you're saying is that, in order for your love for your brother to be real, you have to have a free choice about whether or not to love him. If your love for him weren't freely chosen or if it didn't involve free choices, then it wouldn't be real love. More generally, as you see it, in order for any relationships to count as relationships of love, there has to be free will.

**Rosa:** That's what I think.

**Leif:** What if I could explain that you love your brother because you share DNA, and you were raised by parents who encouraged you to be kind to each other, to share your toys and to spend time together, parents who modeled respect and enjoyment of you and of your brother and of each other, parents who didn't pit you and your brother against each other as competitors for their pride or acceptance, parents who didn't make one of you the "scapegoat" and the other the "golden child," parents who taught you skills for getting along well as siblings and who made fun family vacations and holidays happen, giving you happy memories that you and your brother share and will always have in common. If we took all the factors that are causes of what your relationship with your brother is like, it doesn't seem like we need to cite "free will" at all. There's a full causal explanation for why you love your brother that does not include your free choices.

**Rosa:** Well, that's a scary thought. I don't like the idea that he and I were *made* to love each other or that we had to and had no free choice in it. And I don't mean that we had to in the sense that our parents threatened us with punishment if we didn't love each other – because they never did anything like that – but that the ways our parents acted shaped our relationship so that inevitably my brother and I loved each other. It sounds like your suggestion is that, because of the way things went in our past, there wasn't really any other option. But that just can't completely be right. Because I could choose to be mean to my brother right now – I could call him and cancel the plans we have and say something cruel, so I could harm our love or even destroy it if I chose to. And I make a free choice not to do that. I freely choose to still love him and to treat him in loving ways. That's up to me, not due to my genes or due to my parents.

**Leif:** I wonder if maybe we are getting off track with respect to whether or not love itself conceptually requires free will, in order to count as love.

**Rosa:** You questioned a few minutes ago the idea that love involves free will because we started talking about passion, which I grant seems like a passive feeling. If love were *just* a passive feeling, then it would not require free will to exist. But the other aspects of love we have been talking about, such as desires, actions, and commitments, involve free choices.

**Leif:** Tell me more about how you see the desires that are part of love as requiring free will.

**Rosa:** Okay, back to my boyfriend. When he first started talking with me about the possibility of his applying to a study abroad program in Italy, I didn't want him to go. At first, I didn't even want him to try to study abroad by going through the application process. I was selfishly wanting him to stay on campus here with me, so that I wouldn't miss him for months and be alone here without him. But then, out of love for him, I decided to want him to apply and to want him to be accepted. I thought about how good it would be for him to have that unique experience to travel in Europe and to improve his Italian language skills, and I thought about how much his life would expand by getting opportunities to meet new people in another part of the world. My natural desire was the selfish one of wanting him to stay here with me, but my love for him involved deciding to want him to get the chance to study abroad, for his sake. I think cases like that show that real love requires free choices about what to desire and about what to do regarding the people we love.

**Leif:** Most people probably would agree that it wouldn't have been loving of you to keep wanting him not to study abroad, even if that was best for him, for the selfish reason of wanting him to stay here to keep you company. But still I can't see how, for your love for him to count as real love for him, it had to involve free will. You don't mean that he, as the object of your love, must have free will for you to love him, but rather that you, the one who loves him, must have free will, right?

**Rosa:** Yes. Objects of love – whether people or causes or things or pets or places – don't have to have free will in order to be loved. It's the one doing the loving who has to have free will in order to love.

**Leif:** I think I'm still not seeing it. Suppose your desire for your boyfriend to study abroad (so that his life can go as well as possible overall) was a desire you had because your parents raised you to be a person who cares about other people living lives that are good for them, and they raised you not to see other people just as tools to getting what you want, such as companionship or help.

**Rosa:** I still had to freely choose to let that desire for him to live his best life win out over my selfish desire for him to stay put, in my overall attitude toward the situation. And I had to freely decide to act on that desire for him to live his best life by giving him encouragement to apply, and by showing enthusiasm for his plans, and then by expressing happiness and support when he later got accepted. And then I freely chose to help him plan his classes in the study abroad program, and I freely chose to spend time talking with him about what weekend side trips he would take while he is away.

**Leif:** I see. You think of your love for him as involving free choices at those junctures: to want most for him to stay put with you here or to want most for him to apply; to encourage him to go or to act in grumpy discouraging ways about the idea; to grow cold and distant when he talked about the possibility of studying abroad or to show support; and to join him in being happy when he got accepted or to show anger or even break up with him over it.

**Rosa:** Yes, that's all correct – I think my love counted as real love in part because I made those choices freely, and if I had not had those options to choose among, it would not have been real love.

**Leif:** It still seems to me like those supportive and encouraging actions of yours count as loving actions, regardless of whether or not you had options to be unsupportive or discouraging at the times you acted in those loving ways. I think your actions count as genuinely loving even if you never had any opportunities to develop a different sort of personality than the one you have, a personality that is kind and supportive and encouraging to people you care about (and to strangers, too, from what I can tell).

**Rosa:** Thank you, that's nice of you to say. But I still think that, for my actions to count as genuinely loving, they had to come from *me*, not from anything outside of me, like my parents or just my genetics and events in my past.

**Leif:** Can you explain what you mean by saying that your actions had to come "from *me*"?

**Rosa:** I mean that I had to be the source of them.

**Leif:** What do you mean by that?

**Rosa:** You know, the causal chain tracing back from the free action has to end up at me, not at some hypnotist who made me act on post-hypnotic suggestion I was unaware of, or all the way back through the events of the past to my conception as a zygote.

**Leif:** What are you saying *you* are, then – something like a soul or an immaterial mind?

**Rosa:** I think so. My love has to come from my mind or soul to be *my* love and to really *be* love.

**Leif:** I don't know what to say about that, because I think we are purely physical beings with complex neurophysiology, which, along with the natural laws that govern how physical things work, explains what we do. But I'll talk more about that another day. To focus on what you're saying: I think it's a little odd to think that there is no real love unless there are immaterial souls. Lots of people believe that they genuinely love their partners and their friends and their children without believing in immaterial souls. I don't see how all those people could be misguided about what love requires. Do you think your view that love requires free will can be supported if you were to think of human beings as complicated physical organisms without a non-physical soul?

**Rosa:** Maybe – if the idea that *I* cause my free loving actions could still make sense on that view of human beings as entirely physical. It probably can, although I don't really know how.

**Leif:** Let's say you are a complex biological organism who can think, reflect, reason, feel, form values, set goals, and want things, even though it's hard to understand fully how our brains and nervous systems enable us to do those things. Now we are trying to figure out if your having the ability to love – and anyone being able to love – requires free will. You've said that it does, and in spelling that out, you've said that, to count as genuine love, your love or your loving desires and loving actions must have you as their source or as their ultimate cause. Genuine love requires free will because it requires that you, and not anything outside of you, are its source.

**Rosa:** That describes what I think, yes. It's easy to see how we are each the source of our loves if our loves come from our souls, which transcend the physical world and are not a product of our genetics and past events in our lives. How else do you think we could be the source of our loves?

**Leif:** There are different ideas about how to understand "the self" in relation to our actions that are self-caused or controlled by the self in the way needed for those actions to count as free. Those theories were also covered in my ethics class.

**Rosa:** There are theories that do not conceive of our "selves" as immaterial souls?

**Leif:** Yes. We studied works by the philosopher Harry Frankfurt, who highlighted an interesting feature of us as persons, which is that we can mentally rise above our own desires and beliefs and, by way of reflection on them, form desires and beliefs about our desires and beliefs. In other words, we can form mental states about our other mental states. For example, say I'm out with friends who order a large pizza and it looks really good, so naturally I want to have a slice. But when I contemplate that desire, I remember that I'm trying to avoid gluten and dairy because they upset my digestive system, so I reject that desire to have a piece of the pizza as something I don't *really* want to have or to act on. Instead, I want the desire to refrain from eating the pizza to be the desire I actually act on. If you think of desires to do things as psychological pushes and pulls toward certain actions (which Frankfurt called "first-order desires"), then you can think of desires about desires (which Frankfurt called "second-order desires"), formed by reflective evaluation on your first-order desires, as representing where *you* stand. More particularly, Frankfurt suggested that your desires about which desires you want to actually lead you to action – a subtype of second-order desires, called "second-order volitions" – are what represent your real self.

**Rosa:** Let me see. If some particular Thursday night, for instance, I want to go out with my friends, but I also want to stay in my room and work on a paper for a class instead, those are competing first-order desires.

**Leif:** Right.

**Rosa:** And Frankfurt would say that my real self is represented by the second-level volition concerning which of the competing first-order desires I want to win out that situation. Since I'm a studious person and I like doing well in my courses, I want to want to stay in to work on the paper, and I want that studious desire to be the one I act on rather than the desire to go out.

**Leif:** Exactly.

**Rosa:** So this is a picture of "who I am" that does not refer to an immaterial soul. I am, centrally, my second-order volitions.

**Leif:** Yes. Love, for Frankfurt, is a kind of volitional necessity in which you care about someone in a way that is wholehearted, so that your desires line up at the first- and second-orders in favor of wanting the best for the ones you love and wanting to be with them, and wanting to want the best for your loved ones and wanting to want to be with them. Frankfurt describes love as a matter of caring such that you cannot help caring, which involves desires for the good of the beloved and desires for those desires to be effective in action, so that they count as second-order volitions. On Frankfurt's view, acting freely involves acting with harmony in the psyche, so that you do what you want to do and what you want to want to do. This does not require the ability to do otherwise than what you do at the time, holding fixed everything about you and everything about the past. And so love does not require any kind of forking pathways into the future like we were talking about earlier.

**Rosa:** That's interesting.

**Leif:** There was an intriguing counterpoint to Frankfurt's account suggested by the philosopher Gary Watson, who gave a theory of acting freely as acting on your values. Watson understands values as judgments about what it would be good to do in the circumstances, judgments that you can defend to yourself and others in a cool and non-self-deceptive moment.

**Rosa:** On Watson's view, then, we do not have to have an immaterial soul in order to love or to act freely – we have to have values.

**Leif:** Right. Each of those philosophers has a proposal for who "you" are as a practical agent: your cares or second-order volitions, for Frankfurt, or your values, for Watson. (Watson later changed his view, although this account I've described has remained influential and is still thought-provoking.)

**Rosa:** I do think my values are central to who I am.

**Leif:** In our previous conversations, we have brought up another theorist whose ideas are relevant. Remember last time we got together, we mentioned that Susan Wolf described love as central to meaning in life. Remember, too, that she also has a prominent theory of human freedom as the ability to act in accordance with the True and the Good. We could interpret her as suggesting that *you* are your intellect and will – your faculties for discerning what is true and what is good and for acting in accordance with those discernments – and when you are able to discern truth and goodness with your intellect and are able to act on that discernment by your will, then you act freely. Acting freely does not require, on her account, having any forking pathways into the future, and neither does loving someone.

**Rosa:** I can't help but thinking, though, that we need to come back to the notion of having genuine options. I really think that love involves selecting one option rather than another at times, like the times I described when I could have responded in one way to my boyfriend's proposal of his plans to study abroad and I also could have responded in a different way. It counted as loving, the way I responded, in part because I could have responded differently at the time.

**Leif:** I think that your actions of giving him encouragement and support for his plans were still loving acts even if you could not have done otherwise at the time you did them (or even at any previous time). Frankfurt's view would say that you responded in a loving way freely, in that you acted as you wanted to act and as you wanted to want to act. Watson's view would say that you responded in a loving way freely in acting from your values. Wolf's view would say that you responded in a loving way freely because, at the time, you were able to act in accord with the True and the Good.

**Rosa:** So you are saying that, even if someone thinks that love requires freedom, they do not need to hold that love requires freedom in the sense of sometimes having alternative options as forking pathways into the future?

**Leif:** Yes. And I don't even think it's clear that love requires freedom in any sense, partly because it is controversial exactly what love is. If love is just a strong passion or feeling that overtakes a person, it doesn't seem to require freedom of any sort at all.

**Rosa:** I see what you're getting at. Well, I'm afraid I need to go work on a painting now, Leif, and I love painting! I'll think about this some more and will also reflect on any other reasons I have for believing in free will. See you next time?

**Leif:** Yes, take care, Rosa! See you soon.

# Day 5: Does the Feeling of Freely Choosing Show that We Have Free Will?

Abstract

*Today Rosa argues that we have free will because it feels like we do. It is part of our experience every day to make choices between one thing and another. For example, Rosa says, when she gets up in the morning, she can decide which shirt in her closet to wear, and she can choose to go to her first class or to skip it and go get breakfast instead, and this is all up to her. Leif suggests that, while it is true that we make choices in the sense that we make up our minds about what to do, the question remains of whether or not we "make up our minds" in a way that involves free will. Through the course of the conversation, they distinguish different accounts of our sense of free will, including Dana Kay Nelkin's account of our sense of freedom and, by contrast, a competing sense that requires the availability to an agent of more than one pathway into the future.*

**Rosa:** Hi, Leif! Nice day outside, isn't it?

**Leif:** Sure is! Glad to see you, Rosa.

**Rosa:** You, too. Let's see…today will be my fifth attempt at supporting the claim that we do have free will. Here's my last argument for you. I've been thinking about it, and I've realized I think we have free will because it feels like we do. What I mean is, it is part of our experience every day to make choices between one thing and another. When I get up in the morning, for instance,

DOI: 10.4324/9781003684268-7

I can tell from the inside that I decide which shirt in my closet to wear, whether that's the white one or the black one or the blue one. I can choose to go to my first class or to skip it and go get breakfast instead. Or I can choose to roll over and go back to sleep. It's up to me, and that's just clear in my experience.

**Leif:** So, you think we must be free agents because it seems like we are?

**Rosa:** Well, yes – we have talked about the fact that we make choices all the time.

**Leif:** We do make "choices," I agree, in the sense that it feels like we make up our minds about what to do. But as I see it, what we've been trying to figure out is the answer to the question, do we "make up our minds" in a way that involves free will? Are the choices we make free ones?

**Rosa:** I think that to make a choice itself is a process of weighing options between one thing and another and deciding what to do on our own. We can tell when we do that in the absence of anyone making us choose one way or another. That's free will. It just seems obvious, based on our inner experience, that have free choice. I did some reading last night and saw the term "phenomenology" used to refer to our inner experience or what it is like to be us. Our phenomenological experience as individual people shows us that we have free will.

**Leif:** You're not saying that we act with our own free will all the time, are you?

**Rosa:** No…I mean, maybe if someone with a gun says to my mom that they'll kill me if she doesn't do what they say, and she does what they say to save my life, that's not something that she does of her own free will. She was forced to do something, so as to save her own daughter's life. But that would not feel like a free choice to her. She could *tell* that someone else was making her do what they wanted her to do, by threatening the life of someone she loves. In that case, she does do what they say. But not freely. And she can tell that by what the choice feels like – in that case, it feels forced. Other times, though, when she does things for me, like when she bakes me a birthday cake or buys me a new necklace, she does that of her own free will, which she has evidence for from the inside. She feels those actions of baking the cake and buying the gift as coming from her own free will.

**Leif:** Let me see if I understand accurately. You're saying that, if we're coerced into making the choice we do, like by a severe and credible threat, or if we are subjected to undue force of some kind, then we don't choose freely. But in other instances, we do choose freely. And you think we can tell when we're being coerced or forced and when we're not. When we're not, you see us as having evidence that we are free in making our choices: we can feel that we are and can tell that no one is exerting force over us.

**Rosa:** Yes, that's part of it. I think coercion by severe threats undermines free will, but there are other kinds of cases when we feel unfree, like when we want something extremely badly and that strong desire for it is pathological – such as if it comes from some kind of mental illness, like severe depression, bipolar disorder, or schizophrenia, or from an addiction.

**Leif:** I guess, then, you think we can tell from the inside when a desire is irresistible by us. And you think that, when it is irresistible and we act from an irresistible desire, we do not act freely.

**Rosa:** That's right. I also think that sometimes we feel ourselves from the inside to be weighing among options as we make free choices, and from the internal perspective, we view the future as open-ended. As I've alluded to in our other conversations, I see my own position as a person with respect to my future as a matter of my coming sometimes to forks in the road along the pathway of my life. From lots of conversations with people over time, I get the sense that lots of us think of ourselves as people who at times have forking pathways in front of us, and it *feels* this way, too: it seems that, at least sometimes, it is up to us which one of those forking paths becomes the actual future, because we make it happen by deciding and acting in the way we do of our own free will.

**Leif:** Could you say more about that?

**Rosa:** I gave the example a few minutes ago about freely deciding whether to go to class or to skip it. I also think that I freely decided to declare art as my major, even though my parents were worried about my employability in the future. The decision felt like mine, like one that I made of my own free will. I did not feel an irresistible desire to declare an art major, and my parents did not manipulate or coerce me to do that (in fact, they would have preferred if I had chosen finance or mathematics). It felt as if I was able at that time to choose art as my major, and at that time I also could have chosen math, and also could have chosen finance, and when I chose, I felt as if I did that freely, of my own accord – not from coercion or force and not out of a compulsion.

**Leif:** I see. I guess I used to think of my future like that, and I used to feel like that. But I started to get doubtful when I thought about the scientific evidence, which I'll talk about another day, when it's my turn to try to make my case against our having free will. I heard this analogy that really started to bother me, an analogy which seems to indicate that it is pathetic of us to believe that we have free will, on the basis of our feeling like we have free will. The analogy is this: when we think we have free will, we are akin to a falling leaf blowing the breeze, saying to itself, "now I'll go this way…now I'll go that way…" as if it had any control over where it went at all! And that got stuck in my head. I started to see us as like that: just pushed around by forces we don't always see or feel or fully understand, so that our sense of being free is just an illusion. I don't want to be foolish like a leaf in the wind imagining it's in charge!

**Rosa:** (Laughs.) Okay, well, leaves can't think, and they can't make decisions based on their reasons or values or desires or goals. We are a lot more complicated than that. We can direct the course of our lives ourselves – not completely, of course, since things happen to us sometimes that are out of control, like when my brother got a spine injury that has changed the course of his life. He is in pain every day, can only walk slowly with a cane, and has to spend most of his time lying in bed. That injury was obviously not something that was in his control. But when he and I talk about life, he tells me that he has the sense, too, that he still has free choices to make. He has choices, for instance, about what attitudes to take toward his situation, as well as choices about how to spend his time. He could choose to give up and not try to do anything useful with his mind. But, instead, he studies computer science and game development on his own, by his own free choice, and he feels like there are still some open pathways in front of him going into the future.

**Leif:** I'm sorry that happened to him. That's horrible.

**Rosa:** Thanks. Yes, obviously he's not alone. So many people become ill and injured. The first day we talked, you brought up all the suffering in the world as reason to think there is no God, and I acknowledge that really is a hard problem for anyone who believes in God. My point now, anyway, is that I'm not saying that everything that happens in our lives comes from our free choices. But I am convinced that we do have free will and that we sometimes use it, because we can feel ourselves to be free sometimes from an internal perspective.

**Leif:** All right, I should think more about the internal perspective and what evidence that gives or does not give me.

If I try to be introspective and honest and ask myself – *how does it seem to me from the inside, when I think about what to do next, and when I think about what the future is like in relation to me?* – I guess it feels like, sometimes, I act on autopilot, such as when I'm brushing my teeth and taking medications as I'm getting ready in the mornings – sometimes later I can't even remember if I've done those things, because I wasn't really focusing on them or noticing what I was doing. That's probably not the "sense of freedom" you're talking about, even though I don't think I was coerced by anyone or compelled by an irresistible desire while I did those things. Other times, it feels like I need to make a decision about something, such as whether or not to apply for a certain summer internship.

**Rosa:** When you are deciding whether or not to try for the internship, how does it feel?

**Leif:** It feels like I weigh various considerations, including what the other options for spending my time in the summer might be, and how useful the internship would be toward getting a job after graduation, and how much it might pay, or how in the world I'd support myself if it's unpaid, and what I feel most drawn toward doing.

**Rosa:** Doesn't it feel as if those varying options for the future are *open to you* and feel as if it's *up to you* how to decide, as you weigh up and mull over your various considerations?

**Leif:** Well, I'm convinced by what you might call "evidence from the outside" – evidence from biology and neuroscience and social psychology – that there are *not* multiple future pathways in front of me, actually…it's just that I don't know yet, as I live through time until I get there, which imagined future will become the actual one.

**Rosa:** Are you saying that you don't have a sense of yourself as free at all?

**Leif:** I'm trying to introspect, to think about my own thinking and experience, when you ask that question. I guess if I have any sense of freedom at all, it's that I have the sense that what I do is sometimes partially explained by what I want and by what I believe and what I value – what I do sometimes follows my reasoning. But I doubt that I'm ever really "in charge" of what I do next in the way that you say you have in mind when you talk about forking pathways into the future. And I'm not even sure that my conscious reasoning is ever really the accurate explanation for what I end up doing, even though it sometimes feels like it is.

**Rosa:** Last night, when I was thinking about what final case I wanted to give you to support my belief that we have free will, I did some reading about our sense of freedom. I wanted to see what some respected philosophers say about it. I came across a quote from Immanuel Kant that seemed to fit with what I think. According to Kant, "every being which cannot act in any way than under the idea of freedom is for that very reason free from a practical point of view" (Kant, 1785/1981, p. 50). I interpreted that to suggest that we must be free since we act with the idea or sense that we are free.

**Leif:** I think Kant is hard to interpret, and there's probably scholarly disagreement over his views about free will.

**Rosa:** I'm sure you're right. I also came across a statement from the philosopher Thomas Reid about our having free will. Reid wrote, "This natural conviction of our acting freely, which is acknowledged by many who hold the doctrine of necessity, ought to throw the whole burden of proof upon that side…" (Reid, 1788/1983, p. 344). I took this to be saying that anyone who thinks that we never act freely (but only act of necessity) has the burden of proof in the argument you and I are having, because people have a natural conviction of our acting freely. What Reid refers to as our "natural conviction" is what I'm trying to describe and cite as my evidence.

**Leif:** Interesting.

**Rosa:** Here's another quote – philosopher Terrance Horgan writes, "Normally when you do something, you experience yourself as *freely* performing the action, in the sense that it is *up to you* whether or not to perform it. You experience yourself not only as generating the action…but also as generating it in such a manner that you *could have done otherwise*." Horgan calls that last feature of our experience the "core-optionality aspect of agentive phenomenology" (2015, pp. 36–37).

**Leif:** That's a mouthful. I'm afraid I'm still thinking that what you are referring to could be a self-important delusion.

**Rosa:** Well, another philosopher writing on this topic I came across is Dana Kay Nelkin, who came up another day when you and I were talking about moral responsibility and the kind of control we would need over our actions, in order for us to deserve praise or blame for them. Nelkin has written about our sense of freedom.

**Leif:** Is her viewpoint like yours?

**Rosa:** In a way, she seems like someone I would see as an ally, in that she writes this: "The seemingly unshakeable feeling or conviction that we are free presents a powerful challenge to the idea that we are not free" (Nelkin, 2011, p. 117).

**Leif:** That does sound in line with what you are saying today.

**Rosa:** I think what Nelkin goes on to say about our sense of freedom is different, though, from the way I am thinking of my own sense of freedom. It was helpful to read her work, and it did make me see that there could be an alternative way to give an account of our sense of freedom, and maybe hers is a plausible one. I wonder what you'd say about it.

**Leif:** How does she describe our sense of freedom?

**Rosa:** Nelkin says that, when we pay close attention to what we believe as we rationally deliberate over what to do, and when we pay close attention, as well, to what it is we want for ourselves, we see that this belief and this desire concern the ability to recognize and act for good reasons. She writes that, as we deliberate, and in viewing ourselves as free and responsible agents, we are not committed to believing that there are multiple available futures in front of us. Instead, Nelkin suggests that "…insofar as we deliberate about multiple courses of action, we are committed to our own deliberation having the potential to explain why we perform one rather than the other" (Nelkin, 2011, pp. 5–6). In other words, she describes our sense of freedom as involving a desire to be an *explanatory nexus*, so that what we do is explained by our deliberation over reasons.

**Leif:** I'd be curious to read more about her views. It does sound like she thinks that we certainly do have the ability sometimes to recognize good reasons and to act from our recognition of good reasons, and she sees this as showing that we have free will. There is scientific evidence, though, that we do not actually ever act from conscious and good reasons. This is what I was referring to earlier as "evidence from the outside," in opposition to your argument today based on "evidence from the inside."

**Rosa:** All right, well, I'll listen to your evidence "from the outside" starting next time we get together. But what I want to emphasize today is that I think our evidence "from the inside" supports the claim that we actually do have free will. Again, I'm not claiming that we always act freely. I am only saying that we sometimes do act freely, and we have evidence for our power to act freely in our sense of freedom, which it seems to me involves the availability to us of more than one pathway into the future and our ability to make one of those available futures the actual one. I think my case for my side is pretty much done.

**Leif:** Great, you've given me a lot to think about. Could you sum up your reasons for arguing that we do have free will?

**Rosa:** Yes, sure. I have argued, first, that we are held accountable by God for our actions while we are on Earth, so we must have free will in order for that accountability to make sense. On that note, too, I think that, since there's a God and the world does contain many evils, as you stressed, we must have free will, which explains why there are evils like assault and genocide and betrayal. Those aren't God's fault, but ours, due to the free will God gave us as a gift.

**Leif:** At that point, I objected that we shouldn't base a case for the existence of human free will on the claim that God exists, because that is unknown and controversial.

**Rosa:** Right. I think it's reasonable to believe that God exists, and so do billions of other people, but I recognize that many people are skeptical and don't believe in God. So then, second, I argued that we must have free will because it is required for moral responsibility, and we are morally responsible to each other for many of our actions.

**Leif:** On that topic, I questioned whether we really should hold each other morally responsible for any of what we do in the sense of thinking that people *deserve* praise or blame or reward or punishment for what they do. I also questioned whether moral responsibility requires the kind of free will you described, involving having options available to us to act in some way other than as we do act at a time in exactly the same circumstances, or forking pathways into the future extending from the past. Moral responsibility might require something less than that, or something different in the way of control.

**Rosa:** Yes, I remember that. Then, third, I argued that it is part of being human that we have free will, but you objected to that idea, too, and so I pointed out that human life has meaning, which requires our having free will, so we must have free will.

**Leif:** We talked about differing conceptions of the meaning of human life, and meaning in individual lives, and the relations between those conceptions of meaning and freedom.

**Rosa:** My fourth argument was that love requires free will, and we love some people (and projects and animals and objects), so we must have free will.

**Leif:** Yes. I counterargued that love does not require free will, or at least it does not require free will along certain ways of thinking about what free will is.

**Rosa:** Then, finally today, I pointed to our inner experience as evidence that we have free will. Now I think it's your turn to try to convince me of the other side.

**Leif:** All right, I'll get ready to start that conversation. Let's get back in touch in a few days, if that sounds good. Bye for now, Rosa!

# Part II: The Case Against Free Will

# Day 6: Does God's Existence Show that We Do Not Have Free Will?

**Abstract**

*To this point in the dialogue, Rosa has defended the idea that we do have free will. Over the next five conversations, Leif will take his turn at making a case for the opposing position, aiming to show that we do not have free will. As Rosa started out by citing religious beliefs, Leif perhaps somewhat surprisingly begins his case there, too, arguing in the other direction. He explains the challenge to the idea that human beings have free will posed by the trait of God's omniscience, both from supposition of divine foreknowledge – presenting Nelson Pike's argument for the incompatibility of divine foreknowledge and free will – and from the conception of God's knowledge as from timeless eternity, drawing from Aquinas and Boethius. Leif then explains challenges to the claim that we have free will from its tension with God's trait of omnipotence. Leif himself does not think that God does exist, but he argues that, if God did or does exist, then we human beings would not or do not have free will.*

**Leif:** Hi, Rosa! How's it going? I'm glad we're getting together again, and I'm looking forward to convincing you that we do not have free will.

**Rosa:** Yeah, good luck with that, Leif, I don't think you will! But I will try to listen with an open mind.

DOI: 10.4324/9781003684268-9

**Leif:** Terrific, I'm aiming to keep an open mind, too, even though I admit I have a pretty strong opinion about this.

**Rosa:** Okay, where do you want to start in making your case?

**Leif:** Remember how you started out arguing that we have free will by referencing your religious beliefs?

**Rosa:** Yes, of course. Those beliefs are my main foundation for believing we have free will.

**Leif:** Well, I want to start there, too, not because I believe in God – I guess I'd say I'm agnostic about God's existence and tend toward atheism. I want to start there because I think that religious belief of the kind talked about – belief in a perfect being who is omniscient, omnipotent, and perfectly good – actually supports the conclusion that we do not have free will.

**Rosa:** Why do you say that?

**Leif:** I have a few things in mind. Let's start with God's trait of omniscience. You would probably agree to this: saying God is omniscient is saying that God knows every truth there is to be known, and God has no false beliefs. A perfect being would not be mistaken in any thought that being has. Every one of God's beliefs, if there is a God, is true.

**Rosa:** Yes, that seems right to me.

**Leif:** So here is an argument like one defended by the philosopher Nelson Pike, which is designed to show that, if God has perfect knowledge of all truths, then none of us has free will. Suppose, for example, that God knows right now at 10 a.m. this morning that, later tonight at 9 p.m. you will decide to call your mother.

**Rosa:** All right.

**Leif:** If God knows right now that, at 9 p.m. tonight you will decide to call your mother, then it is true that, at 9 p.m. tonight you will decide to call your mother. That's because all of God's beliefs are true. But if it is true right now at 10 a.m. that, later at 9 p.m. tonight you will decide to call your mother, then how could you possibly be free with respect to the decision to call your mother? Pike argues that the decision to call your mother cannot be a free decision of yours, because to suppose that you *freely* decided to call your mother at 9 p.m. tonight is to suppose that you *could have done something different* at that time, instead of calling your mother – for instance, you could have chosen to continue studying then, or you could have decided to call your boyfriend instead, or you could have decided to go to bed early without having a conversation with anyone.

**Rosa:** Why do you think God's existence could not fit with my having an ability to do something different at 9 p.m. tonight?

**Leif:** Here is why. To say that you could have done any of those alternative things, like decide to call your boyfriend instead at 9 p.m. tonight, is to say one of three things, each of which is *absurd*. First option: it could be that you were able to do otherwise than call your mother at 9 p.m. tonight in virtue of it being true that you could have *changed the past*, by changing God's mind back at the time it is right now, 10 a.m. in the morning – remember we said by hypothesis that what God actually knows at 10 a.m. today is that, at 9 p.m. tonight, you will decide to call your mother. Or, as the second option: to say you could have done something different than decide to call your mother at 9 p.m. tonight is to say that you could have made it such that God has a *false belief* right now and is wrong at 10 a.m. about what you are going to do at 9 p.m. tonight. Or, as the third option: to say you could have done something different than decide to call your mother at 9 p.m. tonight is to say that you could have made it such that the being who believed at 10 a.m. that you would call your mother at 9 p.m. is not perfect, since that being is mistaken, and so you could make it be true that God (the perfect being) does not exist. But each of these three alternatives is absurd. You cannot ever change the past. None of us can. The past is over and done with. You cannot make God have a false belief. God, by nature as an omniscient being, knows all truths and has no false beliefs. God is not ever mistaken. And, if God exists, then you cannot make God not exist. You simply do not have that power over a self-existent necessary being.

**Rosa:** Well, I agree that I can't change the past or make God wrong or make God not exist.

**Leif:** The conclusion of the argument, then, is that, if God knows right now at 10 a.m. this morning that at 9 p.m. tonight you, Rosa, will decide to call your mother, then you are not free in deciding at 9 p.m. to call your mother, because you could not have done anything else then. And this argument generalizes to cover every decision every one of us ever makes. If God is omniscient, then God foreknows, even before we are born, everything we will do during our lifetimes. Therefore, none of us ever acts freely!

**Rosa:** That's a lot to take in.

**Leif:** It is. But it makes sense to me.

**Rosa:** One idea I think of, in response, is something I learned in my high school theology class about the philosophers St. Thomas Aquinas and Boethius. If I remember right, they thought of God as eternal in the sense of being timeless or outside of time. If that's right, then God does not know *ahead of time* what anyone is going to freely decide to do. God simply knows from timeless eternity what happens in the created world.

**Leif:** How do you think that helps?

**Rosa:** Well, it undermines Pike's argument, because the argument you just described relies on there being divine *foreknowledge*. Maybe God is omniscient – God knows all truths there are to be known and has no false beliefs – without having foreknowledge.

**Leif:** To me it seems like we could restate the same kind of reasoning we just went through, even if we supposed that God knows from timeless eternity that, at 9 p.m. today, you decide to call your mother.

**Rosa:** I don't think so, because God's knowing that something is true from the perspective of timeless eternity does not imply that God *makes* that be true. Sometimes *we* (not God) freely make something be true, such as when I declare an art major in college and when I decide to call my mother.

**Leif:** But here's the thing: even if God's knowledge of what you do at 9 p.m. today is not an instance of foreknowledge, it is still an instance of God's knowledge, which is true and cannot be changed by anyone or made false by anyone. And even if God's knowledge of what you decide to do at 9 p.m. tonight does not *cause* it to be true, and you make it true by deciding to call your mother, still if God knows from timeless eternity that you decide at 9 p.m. tonight to call your mother, you cannot do anything different from that at 9 p.m. tonight. The fact that you call your mother at 9 p.m. tonight is just as "fixed" by God's knowledge from timeless eternity as it is from God's foreknowledge.

**Rosa:** I'm not sure I see what you're saying.

**Leif:** If God knows from timeless eternity that you declare an art major, then nothing else can happen at the time that you declare the art major than that you declare the art major. What you do is just as fixed – set in stone, in a way, but set in the mind of God – if God knows from timeless eternity what you do, as it is if God knows *ahead of time* what you do. God cannot be wrong. That's true by the nature of what God is. You can't make God wrong, and neither can I. So, whatever it is that God knows from timeless eternity about what happens in our world is exactly what in fact happens, and it could not be any different from how it is, given God's timeless knowledge of all truths.

**Rosa:** Hmm, okay. Let me try a different reply. Suppose I freely call my mother at 9 p.m. tonight. What *explains* my calling her then is that I freely chose to do that, not that God knew I would do that. The explanation goes the other way around: my freely choosing to call her then explains God's knowledge of what I was going to do.

**Leif:** It sounds like you might be expressing an idea some philosophers attribute to Origen of Alexandria. He wrote, "it will not be because God knows that an event will occur that it happens; but, because something is going to take place it is known by God before it happens."

**Rosa:** Yes – I'm trying to express the idea that God's knowledge of my free action depends on what I freely do, whereas what I freely do does not depend on God's knowledge of it. I imagine there's debate about how to understand the "dependence" I'm referring to and the "because" in what Origen says. Still, this might be a way to save both God's omniscience and my free will in calling my mother.

**Leif:** How do you think this reply gets around the argument Pike poses? Are you envisioning that, when you decide to call your mother at 9 p.m. tonight, you could have done something different instead at that time?

**Rosa:** Yes, I am.

**Leif:** But how does this proposed solution avoid the three absurd consequences of supposing you can do otherwise that Pike sets out? Pike argued that, to suppose you could have done otherwise than call your mother at the designated time has the absurd implication that either you could have changed the past (which you cannot do), or you could have made God wrong (which you cannot do), or you could have made God not exist (which you cannot do).

**Rosa:** I guess what I'm thinking is this: to say that I can still freely choose to do something other than call my mother at 9 p.m., such as choose to call my boyfriend instead, is *not* to say that I can, at 9 p.m., change God's past fixed belief about what I will do. Rather, it is to say this: if I had freely chosen to call my boyfriend instead at 9 p.m., then earlier at 10 a.m. today God would not have believed that I would call my mother at 9 p.m., but instead God would have believed at 10 a.m. that I would call my boyfriend at 9 p.m. tonight.

**Leif:** This idea might be giving me a headache.

**Rosa:** Yeah, I agree this issue is hard to think through. But I need to figure it out, because I don't want to believe that, if God exists, then we don't have free will. Remember, you're the one who posed the challenge!

**Leif:** True! Okay, let me think…honestly, I don't think this solution you've just sketched out really gives you the ability to do other than what an omniscient God knows ahead of time you are going to do. Once we get to 9 p.m. tonight, God's foreknowledge (at 10 a.m.) of your phoning your mother (at 9 p.m.) seems like a fixed feature of the past. And what God knows is true – God can't be wrong. So, you really don't have any other option at 9 p.m. than to call your mother.

**Rosa:** I see what you're getting at…maybe I'm feeling a little stumped.

**Leif:** Well, here's an idea for you. Remember when we were talking about moral responsibility, on the second day of your trying to prove to me that we have free will? We talked about how different philosophers have different conceptions of what free will is. Some of them do not think that, in order to act freely in the sense needed to deserve praise or blame when you do something right or wrong, you need to have been able, at the time you acted, to do otherwise than what you did, holding fixed all the natural laws and the past. In other words, some philosophers think that you can still be free in acting even if there are not forking pathways into the future available to you, with it being up to you which of them becomes the actual future.

**Rosa:** Right. For instance, John Martin Fischer says it's enough for being morally responsible for my action if I act on a reasons-responsive mechanism when I do it.

**Leif:** Yes. And Harry Frankfurt says that you act freely when you do what you want to do and what you want to want to do.

**Rosa:** I recall that.

**Leif:** So maybe there's another way to respond to Pike's argument for the incompatibility of God's omniscience and human free will. You could say that acting freely does not require the ability to do otherwise at the time you act, holding everything fixed, including the natural laws and the past. And if it does not require that ability, then to say that you freely call your mother at 9 p.m. tonight does not have any of the absurd consequences that Pike says it does.

**Rosa:** I see your point. I could say that I freely call my mother at 9 p.m. tonight – even though God foreknew that I would do that – since when I call her, I want to call her and I want to want to do that. Or I could say that I act freely in calling her since I act on a reasons-responsive mechanism. But I'm worried that taking that type of route might leave me stuck in trying to explain some things I believe, such as how some people end up in heaven and some end up in hell, yet God remains perfectly good and fair.

**Leif:** Why do you say that?

**Rosa:** Because if there are never any forking pathways into the future in front of any of us, such that it is up to us which of them to make the actual future, then a person's "pathway of life" is just a straight and unforked road to heaven or to hell, from the moment they are born. And that doesn't seem fair.

**Leif:** Ah, I see. So you want to preserve what we talked about as "genuine options" in how you conceive of free will.

**Rosa:** I think so. Let me try something else. What if I respond to your challenge about God's omniscience and our free will in this way. I'll hold that free will does require our having genuine alternative options for which future to make the actual one when we choose freely. Maybe God knows all truths, but there are no truths about what we are going to freely choose to do until we make those free choices. If that's the case, then there is no divine foreknowledge of our free choices, and there is also no divine knowledge of our free choices from timeless eternity, *until* we make those decisions. On that way of looking at things, it seems that God can still exist as an omniscient being, along with our having free will.

**Leif:** I see what you're suggesting. If that's true, though, then there are things God does not know right now, such as whether or not you will get married and, if so, to whom, supposing you think those are free choices you might make in the future. And if you think that all or most or even some human beings have free will, and you think that we are able to make free choices pretty often or even just sometimes, then that power of ours would leave a lot of the future open and *unknown* by God.

**Rosa:** That does seem odd, that God would be ignorant or in the dark about some things.

**Leif:** Yes, and it would seem to make it hard for God to be in charge of how things go in our world and hard for God to bring about God's plans and aims for our world. That idea you suggested would seem to imply that God cannot providently guide and control what happens on Earth.

**Rosa:** Still, on the idea that God doesn't know what we freely choose to do until we make those free choices, there would not be existing truths that are outside the scope of God's knowledge. There would just be facts that

God comes to know when they become facts. Using our example from a minute ago, maybe it is neither true nor false right now at 10 a.m. that, at 9 p.m. tonight, I decide to call my mother. If I do freely decide to call my mother at 9 p.m. tonight, then God comes to know that truth when it happens.

**Leif:** I get it. But on that picture, God *learns* new truths. God must wait to find some things out until they happen. Some religious believers might not like that idea.

**Rosa:** Why not?

**Leif:** Well, it seems to imply that God changes through time as God acquires new knowledge, and many people think of God as a being that, unlike us, does not change through time. Also, on this line of thinking you are describing, it seems like maybe God isn't perfect, after all, since there are some things God has yet to learn.

**Rosa:** Hmm, I'll have to think about that more.

**Leif:** Let's turn to talking about the divine trait of omnipotence. I think it also poses a problem for the idea that we have free will. God's omnipotence indicates that God is in charge of, and has power over, everything that happens. And if that's so, then *we* are not in charge of anything that happens. To say that God has omnipotence means that God can do anything that is logically possible and that is consistent with God's nature. Changing water into wine is not logically impossible and is not inconsistent with God's nature – with the caveat that we are not talking about a case in which changing water to wine would be wrong and therefore inconsistent with God's perfect goodness (such as the water's being changed to wine for a thirsty man who is deathly allergic to wine). Changing water into wine is something God can do. By contrast, making one thing both taller than, and shorter than, itself is not something that is logically possible, and so it is not something to be done. But turning water to wine can happen without violating laws of logic. Turning water to wine only takes violating laws of nature in the physical world, which God's omnipotence implies that God can do, causing miracles.

**Rosa:** Tell me what you think this has to do with human free will.

**Leif:** Well, since God can do anything that God wants to do that is logically possible and consistent with God's nature, nothing that happens in our world is in conflict with what God wants – nothing is outside the scope of God's will. Everything that happens is subject to God's will and is not a result of our free will.

**Rosa:** Hold on. I think that God's omnipotence, perfect goodness, and omniscience might motivate God to leave some things in the world up to us. God could prevent us from acting freely, and God could intervene and violate laws of nature to prevent the worst consequences of our bad free choices. But we cannot know that, if God were real, God would do those things, simply by knowing that, if God exists, then God is omnipotent, omniscient, and perfectly good. God can *be* all powerful while not always *using* God's power to control everything that happens. Even as an omnipotent being, God can choose to make a world that has free will in it, and that's what I think God did.

**Leif:** But then are you saying that things can happen in our world that are against the will of God? How could anything happen that is outside the control of the will of God, if God is all powerful and in control of everything? It sounds like you think God allows some things to happen that God does not cause and does not really want to happen, like someone's getting viciously attacked by another person. But God must want those things to happen, or God must at least want things like that to happen, for some good reason, or else God is not both omnipotent and perfectly good.

**Rosa:** We talked about this to some extent in our earlier conversation about God and human freedom. I think that God's trait of omnipotence does not imply that God always exerts all the power God possesses to control the way things go, and so I think God can exist along with human free will.

**Leif:** I'm sure there's more to think about on this topic, but to me it seems that the existence of a God who is both omniscient and omnipotent would rule out our having free will, for the reasons I described. I see, though, that I haven't convinced you today that we do not have free will, on your assumption that there's a God. Want to meet up again tomorrow?

**Rosa:** Yes, I'm looking forward to it. How about the same time? See you then, Leif!

# Day 7: Does Causal Determinism Show that We Do Not Have Free Will?

Abstract

*In today's conversation, Leif presents the challenge to the existence of free will posed by the thesis of causal determinism. He describes the notions of physical possibility, the laws of nature, and causal necessitation. Leif presents arguments for the conclusion that the truth of causal determinism would rule out our having free will, including a manipulation argument and a different argument due to Peter van Inwagen. Leif does not claim to know that causal determinism is true, but he explains why he tends to think it is, and the matter is left open. As Rosa and Leif talk, we are reminded of the view of libertarianism and become familiar with the positions of incompatibilism, compatibilism, and hard determinism.*

**Rosa:** Hey, Leif, glad to see you. Let's sit over at this table in the shade today.

**Leif:** Sounds good…This is a nice spot. Okay, let's set aside religious beliefs once again. Like I said, I don't really believe in God, anyway, so the ideas we talked about last time are not my own reasons for thinking that we do not have free will. I just wanted to point out that I don't think being a theist really settles the issue about whether or not we have free will. A person could believe in God and also believe that we human beings do not have free will, from my perspective. They could believe that God controls everything that happens in this world and that

DOI: 10.4324/9781003684268-10

our job is to accept what happens with humility, gratitude, and equanimity, even when bad things happen, because we accept that God makes everything happen for reasons we may not fully understand.

**Rosa:** Yes, I see, and I will give that some more thought. So now, tell me, what is your main reason for thinking that we do not have free will?

**Leif:** To start, there is this thought prominent in my mind: every time we act in some way or other, we act from motives. We are motivated to act by things like our desires, our impulses, our physical drives, our beliefs, and our emotional needs. When you get up from the couch to fill a glass with water, for example, you might be motivated by the drive of thirst. You're thirsty, which motivates you to go get some water. Or if you have company visiting, you might be motivated by knowing that you offered your guest something to drink and by wanting to make good on your offer when they said they'd love a glass of water.

**Rosa:** Yes, that seems right, we do things for various motives or reasons.

**Leif:** Well, you think that we sometimes act freely. But it seems to me that, if we were to ever act freely, we would have to be free to have the motives we have. I mean, for our actions to be up to us in the way free will believers think they are, the motives we have for acting as we do would have to be up to us. They would have to be motives of ours that we have adopted freely. But we don't adopt our motives freely. We just have the motives we do. I can't help that I get thirsty when I get thirsty, and when I do, that thirst motivates me to seek something to drink.

**Rosa:** But we don't just act on biological drives all the time.

**Leif:** Even when we do something because we want to – something like go spend time with a friend who is feeling down – we want that because of how we were raised, because we were taught to care about other people's feelings and to help them when we can, especially our friends. We believe that's how we ought to act because of how we were socialized by our families and by other people around us. None of the motives I have are ones that are *really up to me* whether or not I have them. They have causes beyond me. So, it doesn't seem to be true that how I act is ever really "up to me" or due to "free will." What I do is just driven by my motives, which I have because of how I've been molded to be by various external pressures, biological factors, and other people.

**Rosa:** I see what you're saying. But don't you think we're sometimes free to act or not act on a certain motive? For example, say I am talking to a friend who is feeling down, and I get thirsty. The drive of thirst doesn't *force* me to get up in the middle of the conversation to get something to drink. I can choose to sit still and listen to my friend, deciding that I'll get something to drink later, even though I'm thirsty.

**Leif:** Yes, we do that sometimes, putting our drives "on hold" to satisfy later on. But when you stay there listening to your friend, instead of getting up to get something to drink right in the middle of what your friend is saying, you do that – stay still and continue listening – for some reasons or motives. Maybe it's because you believe it's important not to interrupt your friend when she's saying something significant, and you want your friend to know she's important to you. You act in the way you do because of that belief and that desire. Those are your motives.

**Rosa:** Why should that make me think we don't have free will?

**Leif:** Because the motives you have for acting the way you do are *not* motives you adopted on your own of your own free will. We all have the motives we do because of how we've been shaped to be. We get shaped into who we are, with the motives we have, by factors that are out of our control, factors like our genetics and our experiences in the past, going back to how we were treated as infants.

**Rosa:** It sounds like you're saying you think I have no choice over what kind of person I am and what kind of person I become. I don't think that's true. I think I freely shape the character and personality I have by way of my actions, some of which I think are up to me and truly free.

**Leif:** But in every one of your actions, you are acting on some motive that you already had. You can't create yourself from nothing. You are who you are because of what has happened to you as you've grown up and because of the genetics you have. Let me put it another way. You said that free will and moral responsibility are connected, right. You think of free will as the kind of control over our actions that we would need in order to be morally responsible for what we do.

**Rosa:** Right.

**Leif:** Here's another way to put what I'm trying to say. *What you do* is a result of *who you are*. In order to be responsible for what you do, you'd have to be responsible for who you are. But you are *not* responsible for who you are. You didn't create yourself. So, you can't be responsible for what you do.

**Rosa:** Hmm. I don't think that's right. Maybe there are some aspects of myself I am not responsible for and have no control over, like the fact that my hair is naturally dark brown. There's a tendency to develop diabetes in my family, and many of my relatives have high cholesterol. I understand that I have these genetic dispositions, and you might say those are features of who I am. But I think I can freely choose to eat a diet that is recommended for lowering cholesterol and for lowering chances of developing diabetes. I take a lot of care in what I choose to eat, watching out for good nutrition, and I think those choices are exercises of free will. Even though I am motivated to eat what I do by a desire to avoid getting diabetes and the belief that the meal I am eating is appropriate along those lines, still I don't think just the fact of my having those motives for acting the way I do shows that my act is not free.

**Leif:** But I would say that certain people you've talked with, along with your past experiences, inculcated in you the information about healthy eating and the desires and beliefs you are describing. Those experiences and the lessons taught to you by the people around you *shaped you* into being the kind of person who cares about avoiding those particular health conditions. Let me try to put what I'm thinking in clearer terms.

**Rosa:** Sounds good.

**Leif:** There's a viewpoint called "causal determinism." It says that everything that happens is causally necessitated by past events and the laws of nature. In other words, what happens at every moment is the only thing that could have happened at that moment, given the past and the natural laws that govern our universe (such as the law of gravity). By "the only thing that could have happened" I mean the only event that was "physically possible" at the time, which is the only event that was, at the time, consistent with past events and the laws of nature.

**Rosa:** The causal determinist thinks that the happenings in the universe just sort of unfold from the prior states of the universe, in a straight path without any other alternatives? If I were a determinist, I would think that my life goes along something like a straight train track that has no junctions in it, no forking pathways, so it goes along the only way it can go?

**Leif:** Yes. The causal determinist thinks that every event is causally necessitated by past events and the natural laws. And every action is an event – it's something that happens at a time. So if determinism is true, then every action is causally necessitated by the past and natural laws, meaning that every action *could not have been otherwise* than exactly what it was, given what has come before up to that time and what the laws of nature are.

**Rosa:** And you think determinism is true?

**Leif:** I'm not positive that it is true. But it seems like there is pretty good reason to think that it is. Scientists study the behavior of physical things, including human beings, and they describe the laws that govern the ways they behave. The more we learn in physics, biology, chemistry, psychology, sociology, and neuroscience, for instance, about how the world works, including our bodies and brains, the more we're able to explain why things happen the way they do and the more we're able to predict what will happen next.

**Rosa:** Could you say more about that?

**Leif:** Well, our genetics lay down a blueprint for a lot of what happens in our lives, including our resiliency when we face stressors, our susceptibility to substance use disorders, and what our natural intellectual strengths and weaknesses are. We recognize that some people are "wired" to have a hair-trigger temper or to have ADHD. Researchers also have learned a lot, and they continue to learn more, about the ways formative experiences in the early stages of life shape a person's later behavior. Suffering physical and emotional abuse in childhood has certain predictable effects in later life. What someone experiences in infancy, toddlerhood, in elementary school, on the playground, in the home, and in places throughout the entire course of their past, leads by way of the laws of nature to what happens later in life, including how that person makes decisions and how they act. I want to talk more about evidence against free will specifically from neuroscience and social psychology on other days. But today

I want to emphasize the threat to our having free will posed more generally by the idea that causal determinism is true.

**Rosa:** Okay, but it seems that we don't know whether or not causal determinism is true. For one thing, you mentioned physics, and don't we know from physicists that there is indeterminism at the quantum level? Even if we set aside that idea of quantum-level indeterminacy for today, could you tell me why you think that, if causal determinism is true, then we do not have free will?

**Leif:** Yes. There are different ways to argue for that view. Let me say, first, that the view is called *incompatibilism*: it is the position that, if we have free will, then causal determinism must be false, and if causal determinism is true, then we must not have free will. On the other hand, people who think that free will and causal determinism are conceptually consistent – in other words, people who think both the claim that we have free will and the thesis of determinism can be true at once – are *compatibilists*. According to incompatibilists, compatibilists are mistaken: the two claims – that causal determinism is true and that we have free will – are not compatible with each other; it cannot be true both that we have free will and that the thesis of causal determinism in fact characterizes our world.

**Rosa:** Why do incompatibilists think they are right?

**Leif:** There are various arguments for incompatibilism. Let me start out by telling you a story. Imagine this: you live on a beautiful tropical island called Paradise Island. Your surroundings are gorgeous. There are palm trees waving in the breeze, lining an expansive fine white sand beach by the turquoise water. The homes are small pastel-painted houses with gardens and flowering shrubs. French angelfish and sea turtles swim among the coral reef just offshore. The smell of grilling fish, coconut, and pineapple drift past you. Smiling neighbors pass by and give you a friendly wave.

**Rosa:** Sounds nice so far.

**Leif:** On this island, you are able to do what you want to do, and you find that your every wish is met, including your desires for food, shelter, entertainment, and companionship. However, there are oddities about this island. For one thing, you got to this island not by choice and, little do you know, but you could not leave if you tried. Here's another way in which this island is "fishy" (ha ha). In the restaurant where you eat dinner most nights, one evening you find yourself envisioning some grilled swordfish with papaya for dinner. You order the swordfish, and it arrives at your table. But, unbeknownst to you, the only entrée in the restaurant's kitchen was swordfish with papaya. You could not have done otherwise than have swordfish for dinner that night.

**Rosa:** Interesting.

**Leif:** Do you think that you stay on Paradise Island freely? And do you have swordfish for dinner that night of your own free will?

**Rosa:** I don't think so, but I'm not entirely sure.

**Leif:** Some people will say "yes" to those questions, having in mind an idea of free will roughly as being able to do what you want to do. You want to stay on the island, and you are able to do that. You want to get the swordfish for dinner, and you are able to do that. But there are problems for the viewing free will as simply *the ability to do what you want to do*. We can see that by noticing that the Paradise Island scenario, as I have described it so far, has not yet described any explanations either for why you have the desires you have or for why you act in the ways you do. And we can fill in explanations that undermine an intuition that you are free.

**Rosa:** Like what?

**Leif:** Well, for instance, suppose that in the background of the island society there is a team of neuroscientists who have attached undetectable devices to the brains of all the island's inhabitants. By way of these undetectable devices, the neuroscientists determine the island residents' desires, thoughts, and decisions. In that way, they keep the island society functioning as it does, like clockwork. We can envision this manipulation as completely *hidden* from the island residents and as working in such a way that the residents remain agents, meaning they are beings who desire, reason, form intentions, and act on reasons, but in such a way that the scientists determine what the residents desire and decide.

**Rosa:** Okay, that's hard to envision, but I'll go with your hypothetical scenario for a minute, so as to get your point.

**Leif:** Thanks. So, to continue, on this island, while it is true that you want to remain there, it is also true that it is determined by the scientists that you will always want to stay. When you ordered the swordfish in the restaurant and could not possibly have eaten a different dinner that night in that place, it was also true that your desire for it was determined by the scientists. It was all a set-up. In the morning when you find yourself wanting to ask an attractive person who walks by you on the beach for a date later in the day, that was determined, too: at exactly that time, you would form just that desire regarding precisely that person. And when you see that person, you do exactly what the scientists have determined for you to do: ask that person out for a date. As it turns out, everyone on the island is in the same situation as you are.

**Rosa:** Well, I think that, if I were to get to become *aware* of the "rigged" life on Paradise Island, then I would conclude that, even though I might feel happy about having lots of pleasures in my life, still I do not have free will.

**Leif:** Yeah. And many compatibilists will agree that you are not a free agent on the island as I have described it, with a team of background manipulators. However, they will say, causal determinism is *not like that* – causal determinism merely describes the structure of the natural world, without positing coercive or manipulative agents pulling strings. I, too, agree that manipulators, such as a team of hidden neuroscientists, or other hidden manipulators like hypnotists or puppeteers, undermine the freedom of the decisions they control in other people. But I think that compatibilists who envision the absence of those manipulators, while at the same time thinking that our having free will is consistent with causal determinism, are failing to appreciate what causal determinism means and what it implies about our lives.

**Rosa:** How so?

**Leif:** Well, there are different routes for defending the incompatibilist conclusion, which correspond to different ideas concerning what it *is* about causal determinism that makes it undermine free will. One sort of argument focuses on the idea that causal determination by past events given the natural laws undermines the individual self's being in charge. On this type of argument, causal determinism is *akin to manipulation* by other people who direct the course of someone's life. The central idea is that, if every event – including every decision and action – is causally determined by prior events given the natural laws, then every decision and action springs not ultimately from the self but instead from prior factors that are not the self.

**Rosa:** What are they trying to say?

**Leif:** Causal determinism, *like manipulation*, does not let the person be the ultimate source of her actions: in each instance, her actions are determined by factors beyond her control. This line of argument poses a challenge to compatibilists who accept manipulation as ruling out free will: the challenge is to provide a principled difference between cases in which an agent's decisions and other actions are "merely" causally determined by past events given natural laws, on the one hand, and, on the other hand, cases in which an agent's actions are determined by hidden manipulators, to explain why one acts freely but the other does not. The contention of the incompatibilist is that there is no such principled difference. On this line of argument, if causal determinism is true, then our own lives are actually alarmingly *like* the lives of manipulated people on Paradise Island, in that, in each case – whether we are manipulated or it is "merely" the case that causal determinism is true – we ourselves are not ultimately in charge of what we want and what we decide and what we do.

**Rosa:** That sort of makes sense. Last night, though, I was looking at some of the ideas of the philosopher Daniel Dennett, who says it is pretty silly of incompatibilists to draw analogies between determinism and manipulators like hypnotists and puppeteers, because of what *control* means. To control something is to drive it into whatever state you want it to be in. Hypnotists and other manipulators, like the island neuroscientists, might control other people. But past events and the laws of nature don't *control* anyone, since they don't have any wants or desires about how other people behave. They're not personal, and they don't move anything or anyone around in the world to achieve their own aims, since they don't have any aims.

**Leif:** I understand that point. The past and natural laws don't control anyone in that sense Dennett uses. But still, incompatibilists can say that determinism rules out a person herself being ultimately in charge of what she does, like manipulation rules out a person herself being ultimately in charge of what she does.

**Rosa:** I see.

**Leif:** Let me describe a different argument for incompatibilism that does not refer to its likeness to manipulation, exactly, but one that also gets at the same intuition that determinism rules out any action's being ultimately "up to us." It's called the consequence argument.

**Rosa:** What is the consequence argument?

**Leif:** It's an argument put forward and elaborated by the philosopher Peter van Inwagen. Here it is: It is not right now up to you what happened in the past. It is not right now up to you what the laws of nature are. Determinism's truth would mean that the past and the natural laws necessitate what you do at the next moment. If it is not now up to you what happened in the past, and it is not now up to you what the natural laws are, and it is not now up to you that the past and the natural laws necessitate what you do at the next moment, then it is not now up to you what you do at the next moment. It seems to some people (such as you, I think) that sometimes it *is* up to us what we do at the next moment. But if causal determinism is true, then it is *not* now up to us what we do at the next moment, because what we do at the next moment is the inevitable result of the natural laws and what has come before. Therefore, if we have free will, then causal determinism must be false. And if causal determinism is true, then we do not have free will.

**Rosa:** I think I agree with that.

**Leif:** Peter van Inwagen furthers his argument for incompatibilism by articulating a "no choice principle." The principle is this: if no one has a choice about $p$ (some claim or proposition about the way the world is) and no one has a choice about the fact that $p$ entails $q$ (another claim or proposition about the way the world is), and if $p$ entails $q$, then no one has a choice about $q$. We can take $p$ to be a proposition that expresses the state of the universe at some point in the distant past, even before there were any human beings, together with all the laws of nature. And we can take $q$ to be a proposition stating what you do at a certain time, for instance, that you take the left fork in the trail when you're walking in the woods. If causal determinism is true, then $p$ entails $q$. But if you had no choice about what the universe was like in the distant past and about what the natural laws are, and you had no choice about the fact that these entail your taking the left fork in the trail, then you have no choice about taking the left fork in the trail. If, in fact, you took the left fork and causal determinism is true, then that was the only thing it was physically possible for you to do at the time.

**Rosa:** I don't think we're hemmed in like that. I see what van Inwagen is getting at.

**Leif:** This leads us to a line of argument for incompatibilism different from the "likeness to manipulation" argument: it centers on the idea that, what it is about causal determinism that negates free will is that it rules out genuine options for acting. If causal determinism is true, then at every moment there is *only one way* things can go at the next moment, given what has already happened and absent a miraculous violation of the laws of nature. What happens next must happen, given the past and laws, and nothing other than that can happen.

**Rosa:** So?

**Leif:** Well, many people think that free will requires the ability to do otherwise than act the way you do. We've talked about that. Suppose you freely went left on a trail in the woods at a fork in the trail. If free will does require the ability to do otherwise, then if you freely took the left fork, that would mean that you could have done otherwise than take the left fork in the road. But if determinism is true, then the past and the laws entail your taking the left fork. For you to *do otherwise* than take the left fork (such as take the right fork, or instead turn around and go back the way you came), you would have to be able to make false some truth about the past, or you would have to be able to make false some law of nature. But you can do neither of those. We cannot *now* go back and alter the past. And we cannot now alter (and we could not ever have altered) the natural laws. You have no power over the laws of nature – you can't do anything to make false, for instance, the law of gravitation. The physical laws are what they are. The past is what it is. What comes next in time is unknown by us, but if the thesis of causal determinism is true, then what happens next is, at every moment, fixed by what has come before and by the laws that govern the natural world. So, if determinism is true, you cannot ever do otherwise. So, if determinism is true, you do not ever act freely.

**Rosa:** That makes sense.

**Leif:** To make the idea more vivid, you could think of it this way. Draw a line starting at a point that represents the state of the world at the moment of your birth, and then continue drawing in a straight horizontal line toward the right – moving right represents the passage of time – and stop at a point that represents that state of the world that includes you standing in the woods at the fork in the trail. Now if you are free with respect to which way to go on the trail, it seems that the line from this point – representing the future in front of you – should have (at least) two lines that part ways, making a capital "Y" lying sideways, with the two forking paths in front of the stem, since these depict your ways forward in time, going left or going right along the trail.

**Rosa:** All right, I've drawn that out in a notebook here on my lap. I can see it.

**Leif:** But now we should notice this: the diagram with the sideways "Y" can represent the future accurately only if causal determinism is false. That is because if causal determinism is true, then at every moment there is only *one* physically possible future. That means that, if determinism is true, then for each of us, the diagram representing the path of life as we go through time is a single line with no forks, with no spots along it at which there are genuinely branching pathways into the future. If determinism is true, then – although it may *seem* to us that sometimes there is more than one future available to us – in fact there is not: the mirage of multiple available futures is only that, an illusion. On your diagram, to represent the way things are if determinism is true, you should erase a small part of the right branch of the sideways "Y" where it meets the stem, so that it does not in fact meet. What this illustrates is that you cannot get to that branch from where you are on the stem. Since on determinism, there is *one* future consistent with the natural laws and the past, it would take a *miracle* for you to get to the other (now partially erased) right branch.

**Rosa:** All right, I've erased the connection to the line that angles off to the right.

**Leif:** So, the argument is, if causal determinism is true, then when you ventured left on the trail in the woods, you did not do that freely, despite how it may seem to you. It was as if an invisible (metaphysical) pane of glass kept you from taking the right fork, since taking the left fork was the only thing that it was possible for you to do at the time, consistent with the physical laws and the actual past. If determinism is true, then it would take a miracle for any of us to act otherwise than as we do in fact act. But we need branching pathways into the future – ones that we have the genuine opportunity to take, not ones that would require miracles to reach – in order to be people who are in charge of the direction of our lives by free will.

**Rosa:** Interesting. Let me recap. You think that our past and our physiology dictate what motives we have and so they dictate what we do. You think that every action of ours is causally determined by what has happened to us in the past and by fixed laws of nature that we can't change. And you think that determinism rules out free will for these reasons: one, if it is true, that is too much like being manipulated by hidden other people such as on Paradise Island – there is no good way to draw a principled difference between determination and manipulation – and, two, because determinism rules out there ever being genuine (not illusory) options in front of you for how you can act, since it implies a unique future.

**Leif:** Yes, you've got it.

**Rosa:** I think I'm an incompatibilist who believes in free will. Or maybe I'm a compatibilist. I know for sure that I think we have free will. I guess I'm not fully decided yet on what I think free will is, exactly. But I believe in it.

**Leif:** Here is some more terminology I've read. Someone who believes that determinism is true and also believes that determinism rules out free will, like me, you could call a *hard determinist*. Someone who believes that determinism rules out free will and also believes that we have free will (maybe like you), you could call a *libertarian*. Remember we mentioned libertarianism last week when we were talking about moral responsibility. Also remember, by the way, that it's a different kind of libertarianism than the position in political philosophy. A "metaphysical libertarian" believes we have free will, and free will is incompatible with determinism, so determinism must be false.

**Rosa:** Right. I think I am probably a metaphysical libertarian. (I don't think I'm a political libertarian – I think I'm a Democrat. But anyway…) I'm also attracted to compatibilism, the view that free will is compatible with causal determinism. It's helpful to know about these terms. I need to get going for today, Leif. See you tomorrow?

**Leif:** Yes, sounds great. I want to keep trying to convince you that we don't have free will!

# Day 8: Does Causal Indeterminism Show that We Do Not Have Free Will?

Abstract

*Causal indeterminism is defined as the denial of the thesis of causal determinism. Rosa thinks that causal indeterminism in the natural world leaves room for us to have free will. Leif voices the concern that causal indeterminism would not enable free will but would only provide randomness, and he explains this worry. Non-causal theories and event-causal indeterminist accounts of free will are outlined. Agent-causal indeterminist accounts of free will are also described. The discussion today covers the notions of chance and luck.*

**Leif:** Hello, Rosa! I guess we'd better sit inside today to get out of this rain.

**Rosa:** Sure, let's head into the lobby of this building. There's a nice seating area. (They go in, shaking off their umbrellas at the door, and get settled into some comfortable chairs in a quiet spot.)

**Leif:** All right, so yesterday I talked about why I think causal determinism would rule out free will, and I also said I think that everything we do is determined by factors outside of our control, factors including our genetics and what has happened to us in the past. Today it would be fun to talk about how things look if I'm wrong in suspecting that causal determinism is true.

**Rosa:** Well, I think that would mean that I win!

DOI: 10.4324/9781003684268-11

**Leif:** I want to try to explain why I don't think so. From my perspective, there is good reason to think that, even if determinism is false, we still don't have free will. Let's say we define "causal indeterminism" as the view that denies the thesis of causal determinism. So causal indeterminism is this position: it is not true that, at every moment, there is only one physically possible future.

**Rosa:** If the causal indeterminist is right, then there are sometimes forking pathways into the future, meaning that sometimes there is more than one way things might go, given the way they have gone in the past and given the natural laws that are in place.

**Leif:** Yes.

**Rosa:** And the position of causal indeterminism does not imply that *everything* that happens in the universe, at all times, is undetermined by prior causes. It just means that determinism is false, so sometimes – maybe often, maybe occasionally, maybe rarely – what happens is not determined by what has happened before.

**Leif:** Exactly. Notice, also, that even if the thesis of causal indeterminism is true, that would not show thereby that we have free will, in part because any indeterminism in the world might be only at the quantum level, involving subatomic particles, and that might not help to give us human beings any kind of power of free will at all.

**Rosa:** Why do you say that?

**Leif:** Well, quantum mechanics is complicated and pretty much completely beyond my grasp, I admit. But as far as I can tell we human beings don't gain any control by there being indeterminacy in quantum phenomena. We don't have any control over the particle-like and wave-like behavior of subatomic particles, and we don't have any control over whether some subatomic particle gets reflected or absorbed at some time, and we don't have any control over whether a "wave packet" that physicists talk about collapses in one way or another.

Even if this sort of quantum indeterminacy somehow has effects in our brain activity, that might well not give us any *power* that we could call free will.

**Rosa:** I think that the truth of causal indeterminism might *make room* for us to have free will, if it were the right sort of causal indeterminism and if it were in places relevant to our decision making and our acting.

**Leif:** I'm worried that indeterminism would not give us free will but would only mean that there's some *randomness* in our world.

**Rosa:** Since we last talked, I've read up on some theories of free will that are libertarian. Those are the kind of accounts of free will according to which it is incompatible with determinism. In order for us to have free will as libertarians conceive of it, causal determinism would have to be false and causal indeterminism would have to be true. But the truth of causal indeterminism would only be a necessary condition for our having free will, not a sufficient condition for it. Among libertarian theories, there are non-causal accounts, and event-causal indeterminist accounts, and agent-causal indeterminist accounts of free will.

**Leif:** What is a non-causal account of free will?

**Rosa:** It's a view defended, for instance, by the philosopher Carl Ginet. It says that, at the core of every complex bodily free action, such as raising your clenched fist, is a simple mental act of *volition*, which is itself uncaused. The simple mental act of volition has no internal causal structure, and it counts as an act, Ginet says, in virtue of having an *actish phenomenal quality*, which is the quality of it seeming to the person as if it were directly caused by him (although it was not caused by anything).

**Leif:** Interesting. So that's an example of a non-causal libertarian account. And the question of whether or not we have free will, on this account, would come down to the question of whether or not there are ever any simple mental acts of volition of ours that are uncaused by anything.

**Rosa:** Right. To illustrate the account, your action of casting your vote for a certain candidate in an election counts as free when it is preceded by your uncaused mental act of volition to cast your vote for that candidate.

**Leif:** It doesn't seem all that natural to me to think of our free actions in that way (setting aside that I don't think there ever are any). I mean, it doesn't seem that, when I try to envision what free actions are, I conceive of them as having at their core mental acts that derive from nothing and come out of nowhere. When people say they believe in free will, I'm not sure that's what they're saying they believe in.

**Rosa:** It's true that adopting a non-causal account of free will would require denying the principle of universal causation – the idea that everything has a cause.

**Leif:** Yeah, it's just not intuitive to me to deny that. I'm attached to the idea that everything has a cause, and science seems to be rooted in that idea, too. Maybe even rationality itself assumes the principle of universal causation. But I suppose I can see why someone might think that free will requires something that is totally uncaused.

**Rosa:** Next, there are event-causal indeterminist accounts of free will defended by some philosophers. Those views say that free will does not require denying the principle of universal causation. Instead, they say, we could think that some causes bring about their effects in a way that is probabilistic or non-necessitating rather than deterministic, and we could think that such cases of causation are key to characterizing free actions. Philosophers who defend views like this include Robert Kane, Laura Ekstrom, and Christopher Franklin.

**Leif:** How do those types of theory work?

**Rosa:** Well, there might be indeterministic causation of the right sort and in the right places in the decision process leading to a free action. The right sort might be indeterministic causation among certain events that involve the person herself, and the right places might be between the reasons of the person's own and the agent's subsequent decisions and other actions.

**Leif:** Could you say more about that?

**Rosa:** An event-causal indeterminist account of free will can suggest this: a mental act of decision or a bodily action is directly free – meaning its freedom does not derive from an earlier free action – when that action is caused indeterministically and in the right way by reasons of the person's own, such as her convictions, desires, values, beliefs, intentions, and preferences, and, in addition, the act is not compelled or the result of manipulation, undue force, or coercion. A person's having free will about acting in some way, at some certain time, requires it being open to her at that time not to perform the action, either by doing something different right then or by not doing anything at all right then.

**Leif:** I guess I'd need to read a little more about "indeterministic causation," in order to fully understand that. But I think I get the basic idea: a person's free action can be caused, so long as it is indeterministically or probabilistically caused by reasons the person has for doing it.

**Rosa:** An example would be this: suppose you have a strict deadline for getting a paper finished for a class, and you want to meet this deadline to get a good grade, but you have a conflicting pull on your attention to help your friend who needs your time after his relationship breakup. Your decision to spend the time working on the paper, instead of spending that time helping your friend, counts as a free one (according to event-causal indeterminist accounts) if you weren't compelled or forced to decide in that way *and* your decision was caused in a non-necessitating way by your reasons, such as your belief that good grades are important to your future, your desire to make a good impression on your professor, your valuing timeliness, your belief that you can help your friend tomorrow, and so on.

**Leif:** Hmm, I'm wondering something about this approach. Doesn't it just introduce randomness and accident into the picture of a person's decision process? That is my bottom-line reason for thinking that free will is ruled out by causal indeterminism.

**Rosa:** A defender of the event-causal indeterminist view of free will would say that your free act of working to finish your paper is not *random*, since it is caused and justified by your reasons. It is not an event that appears out of nowhere. It is not some sort of uncaused happening. It is not a miracle. And it is not *accidental*, since it is done on purpose. It is done intentionally and for reasons that favor doing it.

**Leif:** Okay, I'll think more about that. Let's hear about the third kind of theory.

**Rosa:** Sure. There are agent-causal libertarian accounts of free will. This type of account is defended by philosophers including Roderick Chisholm, Meghan Griffith, and Timothy O'Connor. Agent-causal libertarian accounts differ somewhat among each other, but in general they say that when someone acts freely, the person himself causes his free action.

**Leif:** Don't all free will theorists say that?

**Rosa:** Well, yes, but when I say, "José ate the last piece of cake," I might use that as short-hand for what I really think happened, which is that events in José brain, like the brain events that correspond to his feeling hungry, and the mental events of his knowing that I wanted that piece of cake, and his belief that I would not get too mad if he ate it, and his desire to eat it, led in a causal chain of events to the event of his muscles working so as for him to serve himself up the cake and bring it to his mouth and chew it. I would mean that a series of events together made up what makes it true that José ate the last piece of cake.

**Leif:** Okay, and how is that different from what the agent-causal libertarian means?

**Rosa:** It helps to know that in using the word “substance,” philosophers usually don’t mean a white powder or illegal drug. They use the term broadly to refer to a thing, object, or entity. Substances exist rather than happening – as opposed to events, which do happen at times – and they are different from properties, which are traits or attributes.

**Leif:** Got it so far.

**Rosa:** Agent-causal libertarians about free action say that agents themselves, as substances, bring about free actions (or they directly cause some precursor to or part of a free action). And causation by the agent, they think, cannot be reduced to, or fully restated accurately by referring only to, events that involve the agent. They also hold that the agent’s causation of the free act cannot itself be determined by anything. They maintain that the agent had to be able, at the time of the free act, to exert her agent-causal power to bring about the act and also able, at the time of the free act, not to exert her agent-causal power to bring about that act (by not doing anything at that time or by doing something different, instead).

**Leif:** So how would an agent-causal libertarian describe José’s decision to eat the last piece of cake?

**Rosa:** They would say that José decided to eat the piece of cake freely if the thesis of causal determinism is false and Jose himself agent-caused his decision to eat it, without anything outside of José himself determining that he decided that. When José acts freely, he is an “undetermined determiner” or an “unmoved mover.”

**Leif:** Causation of an event, such as a decision, by *an agent as a substance* seems kind of a mysterious and strange sort of causation to me.

**Rosa:** Some people think that substance causation is the main type of causation in the world. If they are right, then maybe thinking that people can "agent cause" their volitions or intentions or actions – in the sense that the agents themselves, as substances, bring about their volitions or intentions or actions – is not so odd.

**Leif:** The way I see it is this: things *happen at times*, and those bring about other things happening at other times. By "things" I don't mean substances. I mean events. Events take place at particular times, such as when the baseball bat strikes the ball at a certain moment. The striking event happened at that time.

**Rosa:** Do you think there could be probabilistic causation among events?

**Leif:** Probably. (Laughs.) Actually, I'm not sure.

**Rosa:** Agent-causal libertarians think their view is better than the event-causal libertarian view because they think it's more plausible to say that when someone does something freely, she makes it happen, rather than saying that events inside of her might or might not lead to its occurrence.

**Leif:** Why?

**Rosa:** Well, one concern is that probabilistic causation in decision-making might seem to make that decision just a matter of chance. However, I do think that if *chance* is defined simply as something happening that had a probability of happening less than 1.0, then that's not really a good criticism of the event-causal libertarian view. Saying that something happened "by chance" could mean just that it had a probabilistic cause. And scientists do seem to think that probabilistic causation happens. Agent-causal libertarians think their view is the best because it straightforwardly captures the idea that free actions are under the person's control. She herself causes them.

**Leif:** You've explained the three different types of libertarian theories of free will – thanks, Rosa. But, you know, I still think free agents – on any of the libertarian accounts of what that means – would just be *lucky* when they make good decisions, and they'd be unlucky when they make bad decisions. If there is indeterminacy in how they decide – if they could decide one way at the time and they also could decide a different way at the time – then the way they end up deciding is just a matter of luck for them.

**Rosa:** Could you say more about what you mean?

**Leif:** We've talked about three types of incompatibilist theories of free will. Someone might say that there is indeterminacy in how free agents decide in that there is *no cause* of how they freely decide. Or someone might say that there is indeterminacy in how free agents decide in that there is *nondeterministic event causation* involved in how they freely decide. Or someone might say that there is indeterminacy in how free agents decide in that there is no deterministic event-cause of their *agent causing* one decision rather than agent causing a different decision. But the way I see it, on all those libertarian pictures of free will, what people decide to do is really just *up to luck*, not up to them.

**Rosa:** I think you have to be careful when you say something is "just up to luck" – because that can make it sound like you are referring to someone who the matter is "up to," like some mysterious force or some being in the universe, like Lady Luck.

**Leif:** I don't believe in that – I don't think that there's some "force" or supernatural person called Lady Luck. I just mean to say that all the libertarian theories you've described seem not to give human agents real control over what they decide and what they do.

**Rosa:** Defenders of all three types of libertarian theories we've talked about can say that their theories are better than compatibilist accounts of free will. They say this because, on the libertarian theories, if we have free will, then sometimes, in the exact circumstances, holding fixed the natural laws, a person can make one decision and can do something other than make that decision. And this is so not just in the sense that the person has the needed skills for making one decision and the skills for making an alternative decision, but also in the sense that the universe leaves room for the agent to exercise these skills, giving her the opportunity to make one decision or another (or no decision at all).

**Leif:** Skills plus opportunities to do otherwise.

**Rosa:** Yes. So incompatibilist accounts secure the sense many of us have of ourselves as free in part because sometimes we can extend the past in one way and we can extend it in an alternate way instead. In front of us in time, sometimes, there are forking pathways into the future. Compatibilist accounts of freedom do not capture this aspect of the sense many of us have of ourselves as free agents and the corresponding attitude we have concerning the openness of the future.

**Leif:** From my perspective, though, if causal determinism is false and causal indeterminism is true, we still don't have free will. I think free will is incompatible with determinism, as I explained last time we got together. And I think it is incompatible with indeterminism, too. Indeterminism I still see as just chaos or chance or luck. It doesn't give us control. So, either way – whether causal determinism is true or causal indeterminism is true – I don't think we have free will.

**Rosa:** I haven't been convinced by what you've said in favor of your idea that causal indeterminism rules out free will. To me it seems like you've really just said it equates to chance or luck or randomness. But I think some of the libertarian accounts of free will sound like they could hold up and make sense.

**Leif:** I'll do some more reading on that, Rosa. But I doubt I'm going to be convinced. Let's move on next time to studies in social psychology that go against the idea that we have free will.

**Rosa:** Sure thing, see you soon, Leif!

# Day 9: Does Social Psychology Show that We Do Not Have Free Will?

Abstract

*Leif has been affected by a professor in his psychology course, who pronounced that human beings do not have free will in view of evidence that people's behavior is driven by factors other than conscious reasons. Experiments in social psychology, Leif argues, show that we should view free will not as something real but instead as an illusion. He cites an experiment showing that fewer people step up to help a victim, the larger the group they think they are in, even though this does not provide a good reason not to help. Leif also describes Philip Zimbardo's Stanford prison experiment, in which the behavior of both study participants assigned as guards and those assigned as prisoners deteriorated rapidly, suggesting that we are strongly influenced by our situations, including the rules we are told to follow, as well as by our desires to conform to social expectations. Leif also refers to Stanley Milgram's famous experiments, which are taken to show the power of a human drive to obey authority and a lack of compassion driven by circumstances. Finally, Leif describes research on politically motivated reasoning. Rosa is not convinced that any of this data rules out free will.*

**Leif:** Hi, Rosa! I'm looking forward to talking today about research in social psychology that supports my position that we do not have free will. You know, when I think about it, studying psychology is what first led to me

DOI: 10.4324/9781003684268-12

accepting that free will is an illusion. In my first psychology course, the professor told us that, even though we might want to imagine ourselves as fully in charge of what we do because it makes us feel important, we do not actually have free will, because there is powerful evidence showing that people's behavior is driven by factors other than conscious reasons.

**Rosa:** I'll be glad to hear the supposed "proof" against what I believe in. (Laughs.)

**Leif:** Okay, well I'll start with a pretty well-documented finding called the bystander effect. You've probably heard of it.

**Rosa:** Yes, I have.

**Leif:** The basic idea is that, when we are in a large group of people who are aware that something is going wrong, we are much less likely to help a person who needs help than we are if we are not in a group. In other words, simply being in a situation in which there are many potential helpers causes us not to take on the burden of helping ourselves. It's like we assume "someone else will do it, it doesn't have to be me."

**Rosa:** What would that have to do with free will?

**Leif:** Well, if the situations we happen to find ourselves in (for instance, being in a large group when something is going wrong) drives what we do (such as, not bothering to help someone in need of help), then that indicates that we do not freely choose what we do in response to events. Our actions are determined by our circumstances.

**Rosa:** Hmm. Let's get into some concrete studies.

**Leif:** Sure. You've probably heard of a famous experiment on bystanders in the wake of a stabbing of a woman named Kitty Genovese in New York City in the mid-1960s. Kitty returned to her apartment complex at 3 a.m., tried to avoid a threatening man, and yelled as he grabbed and then stabbed her. Although some neighbors witnessed the stabbing from their apartment windows, no one went to help her, although one man did yell "Let that girl alone!" according to the New York Times (1964). The assailant seemed to leave, but he came back and stabbed Kitty again – then left when lights in the apartment windows came back on and windows opened, indicating that bystanders were aware of the continued attack. The assailant left again, but then returned yet again and continued to stab Kitty. No one tried to help her, and no one called the police until 3:50 a.m. The police arrived two minutes later to find Kitty already dead. When asked by police why they did not intervene or call the police sooner, witnesses said they didn't want to get involved, or they were afraid, or they were just tired and went back to bed.

**Rosa:** Yes, I've heard of that case. It's awful.

**Leif:** Researchers named John Darley and Bibb Latané in 1968 published the results of a study inspired by the Kitty Genovese case. Participants in the study were each put into separate rooms alone, where they were told they would be speaking into a microphone and listening to other study participants talk about their problems as a university student in a competitive urban environment. There were three groups in the study: in group 1, the study participants were told there was only one other participant in their group; in group 2, they were told there were two other participants in the group; and in group 3, they were told that there were five other participants. So, if student participants were in group 1, they thought they were in a group of two people total. If they were in group 2, they thought they were in a group of three. If they were in group 3, they thought they were in a group of six people.

**Rosa:** Got it. And what happened?

**Leif:** As the researchers set up the study, the supposed "other participants," those who were being listened to, were just recordings. There were not any other real people in any of the study participant's "groups." The study participants were told that their microphones would not work while someone else was talking. They would then, at one point, hear a student say over the microphone that he wasn't feeling well and felt he was about to have a seizure (after he had mentioned earlier his having a seizure disorder). The "victim" (person in distress) would ask for someone to help him, become incoherent, say in halting speech that he was going to die, and then there'd be choking sounds, and the voice would cut off completely after 125 seconds. The researchers wanted to see when, and if, a study participant would leave their own room to go seek help for the person they heard who was in distress.

**Rosa:** The researchers were testing the hypothesis that, the larger a group of potential helpers a person thinks they are in, the less quickly and the less likely they are to help a victim in an emergency.

**Leif:** Exactly. And the study results strongly supported that conclusion. As subjects believed they were in a larger group of people, the percentage who went for help *quickly*, as the seizure was (supposedly) happening, declined dramatically. In group 1, 85% of participants left the room to get help quickly. In group 2, 62% of participants did so. In group 3, only 31% did. Among those who *eventually* helped, the amount of delay in helping was greater, the more bystanders were thought to be present. The total number of subjects who eventually reported the emergency at all, before the experiment was cut off, also varied among the groups in a similar way: all of the participants in group 1 eventually reported the emergency, but only 85% of group 2 did so, and only 60% of group 3 did.

**Rosa:** So, the study supports the idea that people tend to get more passive, and not step up to take responsibility for helping someone, the larger the group of bystanders or witnesses they are in.

**Leif:** Right. Many people seem to like to think that they have free will, and they do not like to think that their behavior is driven by their circumstances. But studies like Darley and Latané's support the conclusion that what we do is shaped by, and is predictable by, what circumstances we are in. Like it or not, this and other social psychology studies show that what we do is not entirely up to us but is instead determined the situations we're in, such as being in a large group of bystanders.

**Rosa:** I agree that it's a disturbing finding about us as human beings.

**Leif:** Let's talk about another couple of famous social psychology studies: Philip Zimbardo's Stanford prison experiment, and Stanley Milgram's studies on obedience to authority.

**Rosa:** Yes, I've heard and read about both, and I refreshed my memory last night. Zimbardo was a psychology professor at Stanford University who managed a research team. I read that his famous experiment in 1971 started with newspaper advertisements seeking male college students for a study on prison life. Some of those who volunteered were picked up by the real Palo Alto police and "arrested" as prisoners. They had not been told in advance this would happen, so it was a surprise, and they were put through a full booking procedure at the Palo Alto Police Department, charged with armed robbery and burglary.

**Leif:** Right.

**Rosa:** Other volunteers were assigned to be prison guards. They were given guard uniforms and were told to keep the prisoners in line and to prevent them from escaping. The point of the study was to examine the effects of situational variables on participants' reactions and behaviors. Zimbardo was interested in how easy it was for ordinary people to do anti-social and dehumanizing things to other people when they were in situations where they felt anonymous or could perceive others as less than fully human, as objects or enemies.

**Leif:** Exactly. What happened was so disturbing that the experiment had to be cut off on the sixth day, even though it had been planned to last for two weeks. The subjects had been chosen by screening assessments, which were supposed to exclude people with medical problems, criminal backgrounds, and psychological impairments and to include volunteers who were psychologically stable and healthy. The roles of prisoner or guard were assigned randomly. The experiment was conducted in the basement of Jordan Hall, the location of Stanford's psychology department, where mock cells were set up, each very small and each holding three prisoners. There was also a corridor for a prison yard, a closet for solitary confinement, and a larger room for the guards and a warden. Guards were given access to other areas for relaxation and were not required to stay on the premises after their shifts. At the guards' orientation session, they were told not to physically harm the prisoners, but they were given wooden batons to show their authority and mirrored sunglasses to prevent eye contact and create anonymity, in addition to their uniforms. They also were instructed to make the prisoners feel as if they were in an actual prison.

**Rosa:** I recall that, after the arrests and being transported to the Stanford "prison" in police cars with sirens wailing, the prisoners were strip searched, made to wear uncomfortable smocks without underwear, and given an identification number, which guards used to refer to them rather than using their names.

**Leif:** That's right. On the second day of the experiment, the prisoners tried to rebel, but the guards reestablished their control by spraying them with fire extinguishers and removing the mattresses from their cells. Over the next few days, guards humiliated prisoners by forcing them to count off and do pushups arbitrarily, keeping them from using the bathroom, and forcing them to relieve themselves in a bucket in their cell. They also divided prisoners against each other by treating them differently based on their "good" or rebellious behaviors – for instance, guards would give back or take away mattresses and clothes. Guards also commanded prisoners to do pointless tasks, like move boxes from one place to another and back again, and some prisoners were made to clean toilets with their bare hands. By the fifth day of the experiment, visitors were allowed, but parents as well as two other psychologists were so disturbed by the changes and behaviors they observed in the study participants – such as extreme emotional distress in prisoners and guards' forcing prisoners to wear bags over their heads – that they called for the experiment to be stopped. Zimbardo ended the experiment the subsequent day.

**Rosa:** I know the experiment has been criticized on many grounds and that ethical guidelines for participation in psychological research experiments were made stricter in the wake of it. But what do you think it shows about human nature and free will?

**Leif:** I think it's a stark illustration of the ways in which our reactions and behaviors are molded by the roles we play in society, as well as by the rules we are told to follow, and by the rewards and punishments that come our way at the hands of other people. Prisoners in the experiment became docile and distressed in response to their circumstances and in response to how the guards and other prisoners treated them. They adopted the vision of themselves that others projected onto them, such as that they were a "bad prisoner." Prison guards adapted quickly to their positions of power and authority, some misusing it to the point of abuse, and they seemed eager to please the designers of the experiment by keeping the prisoners in line as instructed.

**Rosa:** Don't you think the participants in the study retained their free will?

**Leif:** The experiment, from my perspective, supports the conclusion that our situations, including our social roles and the rules that are ingrained in us, shape how we feel about ourselves and how we behave. It makes me, and a lot of other people, question whether we ever really act of our own accord freely, or from some true self that somehow transcends social conditioning and the situations we are thrust into in our lives. Since the study participants were assessed prior to the experiment as psychologically healthy, it makes me wonder whether any of us could behave in the ways the prisoners did, if we were subjected to the arrests and treatment they endured, and if we would behave in the ways the prison guards did, if we went through the orientation session, experiences, and roles they were put in. It seems that it was their situations, rules, and roles that determined how they behaved.

**Rosa:** I do understand that interpretation of what happened, but I'm not sure I agree. Let's talk about Stanley Milgram's experiments about obedience to authority (Milgram 1974). I gather that Milgram wanted to understand why people follow commands to do horrible things they seemingly would not otherwise do on their own, such as guards in Nazi prison camps during the Holocaust.

**Leif:** Right. Stanley Milgram was a Yale University psychologist. In his experiments, study participants were told that there were helping psychologists understand how punishment was related to memory. But really, Milgram was interested in the ways people behaved when put in experimental conditions in which they're told by an authority to do something that would seem to go against their personal conscience.

**Rosa:** I remember that there were roles assigned as "teacher" and "learner" in Milgram's studies. The participant would draw a slip of paper, along with another person (who was in fact secretly in on the experiment and not a volunteer), to see which role they each got. It was arranged so that the study participant was always the teacher. The teacher would then watch the learner get strapped into a chair, which was set up to look as if it was electrified and designed to deliver shocks to the learner. No shocks were ever, in fact, given, but the teacher was told that the shocks were real, and the learner would act as if the shocks were real.

**Leif:** Yes. There were different versions of the experiment over the years Milgram performed them in the 1960s and '70s. In the earlier experiments, after seeing the learner strapped into the electrified chair, the teacher was moved to another room where he could no longer see the learner. The teacher was instructed to deliver a shock to the learner for each wrong answer to a question, by way of levers that were associated with different degrees of shock and arranged into levels. The voltage of the first shock for the first wrong answer was lowest, then the voltage increased for the second wrong answer, and so on. There was a label reading "intense shock" for the levers at a level about half-way through the sequence (the last of those labeled 300 volts), then a label reading "extreme intensity shock" for the levers at the next level, then "dangerous shock" for the levers at the next level, then finally "XXX." The learner answered questions by pressing a button. Teachers were told to report aloud the voltage of the shock they were about to deliver to the learner for each wrong answer, starting from 15 volts and going up to 450 volts at the end.

**Rosa:** I've read that, even while the teachers in the experiment behaved in some anxious ways, such as twitching or nervously laughing or sweating, whenever they brought up the possibility of stopping, the experimenter would calmly reply, "Please continue," or "The experiment requires that you continue," or "It is absolutely essential that you continue." If the study participant continued to object and adamantly said they wanted to stop, the experimenter replied, "You have no other choice, you must go on."

**Leif:** That's right. Experimenters told the study participants that a learner's failure to respond at all should be treated as a wrong answer. At the point after getting a twentieth shock (which was at the "intense shock" level), the learner would pound on the wall, and then the learner did not answer any more questions after that. The study result was this: none of the teachers stopped giving shocks before the twentieth shock. Five study participants stopped after that twentieth shock. Four participants continued to the first lever in the level labeled "extreme intensity shock." Four stopped somewhat later. Twenty-six of the forty participants kept giving shocks to the learners all the way to the end, up to 450 volts.

**Rosa:** Good grief. In later versions of the experiment, some details were changed.

**Leif:** Yes. In one version, the teacher could hear the learner speak, and so the teacher heard grunts in response to the 75-volt shocks, and they heard the learner say the shocks were becoming painful at the 120 volt "moderate" level. Then came groans from the learner, refusals to continue, then screams that he cannot stand it at 180 volts, then screams of agony at 270 volts. Violent screams and screeches in agony continued with increasing severity of shocks. Out of the forty study participants, twenty-five kept going and delivered shocks all the way to the end to the very highest level.

**Rosa:** It's really disturbing to hear what people will do when someone in authority tells them what they must do and says that they have no choice.

**Leif:** I agree! There was even a version of the experiment in which the teacher would have to force the learner's hand back onto a plate that was supposedly electrified in order to continue to shock him. And study participants kept doing it. In that version of the study, twelve participants did it over and over all the way to the highest voltage.

**Rosa:** At least that's not everyone who did that, but it's still pretty shocking (no pun intended).

**Leif:** Yes. To me, Milgram's experiments are a vivid illustration of the way people are quite predictable and tend to do what they're told to do. They obey the rules they are taught. And even though the rest of us are not in those experimental conditions of delivering shocks to learners, still every one of us is told how to behave from the time we are little children. We are given rules to follow, from authority figures in our families and in society. We might not even remember all the rules that have been subtly ingrained in us throughout our upbringing and social conditioning. But still we behave according to those rules. We do as we have been told.

**Rosa:** Now that we've talked about some famous social psychology experiments, what do you think that we should conclude from them?

**Leif:** I think, taken together, they show that we human beings have very little control over our behavior, maybe even no real control by anything you'd call "free will."

**Rosa:** Why do you say that?

**Leif:** Well, the experiments indicate that human behavior is driven by the situations we find ourselves in and by the effects those various situations have on our unconscious automatic behavior-producing processes. The situations and the effects of them leave little to no room for our conscious reasoning or "true self" to be ultimately in charge of how we behave.

**Rosa:** I hate to admit it, Leif, but you are succeeding in making me doubt that I have some "true self" that can completely transcend all my social conditioning, social roles, and the rules that have been ingrained in me about ways I should behave. Gender norms, for instance, I'm sure have affected aspects of my personality and the ways I interact with people and present myself. I agree that our situations indeed do affect what we think, how we feel, and what we do. But I'm not sure if that means I have no free will at all.

**Leif:** Let me add a little more to my case by bringing up some studies on politically motivated reasoning.

**Rosa:** Sure.

**Leif:** Motivated reasoning happens when we do not look at evidence in a completely neutral, objective way, but instead (unconsciously) cherry-pick it to support some conclusion we want to reach. For example, think about someone who wants to leave his wife so he can start a new life with someone else, but he doesn't want to be at fault or take responsibility for causing a divorce. He might unconsciously project desires for infidelity onto his wife and look for evidence that she wants to leave him or evidence that she is guilty of cheating on him. The wife's friend who vouches for where his wife was when she was out on a certain night, for instance, the husband will think must be lying, and the little heart his wife marked on a certain day on her calendar must indicate that she got together with a lover that day, in the husband's mind. He construes evidence so that it "supports" the conclusion he wants to reach. The (unconscious) goal of his reasoning about his wife's behavior and attitudes is to reach the conclusion that she is somehow guilty, so that he can justify leaving her.

**Rosa:** I understand the idea. What interests you about motivated reasoning in the realm of politics?

**Leif:** Several studies show that the unconscious goal we have when we are reasoning about politics is not to find out the truth, but instead to protect our sense of identity as a member of a certain group, such as Democrats or Republicans. We want to stay part of the team, like an avid sports fan, and we want to see evidence that supports the conclusion that our team is the best team and the right team.

**Rosa:** I have noticed that in everyday life, honestly.

**Leif:** Here's an interesting study that shows how people's loyalty to "their side" in politics shapes how they interpret information. It was published in the *Stanford Law Review* in 2012 by Dan Kahan and his co-authors. The study subjects all watched a video of a protest. Half of the study participants were told that they were watching a protest at an abortion clinic, and the other half were told that they were watching a protest against a military policy banning openly LGBTQ Americans from military service.

**Rosa:** Interesting idea for a study. So, the study participants didn't actually know what kind of protest they were watching video footage of. They were just told it was one kind or the other.

**Leif:** Exactly. Participants were then asked questions about what they had seen. For instance, they were asked whether or not the police on the scene controlling the crowd had violated the protestors' rights. Among participants who identified themselves as political conservatives, 70% said that the police violated protestors' rights when they thought they had watched a protest at an abortion clinic, whereas only 16% of them said that the police violated protestors' rights when they thought they had watched a protest of the military LGBTQ policy. On the other side, among those who identified themselves as political liberals, 76% said that the police violated protestors' rights when they thought they had watched a protest of the military policy, while only 28% said that when they thought they had watched a protest at an abortion clinic.

**Rosa:** The subjects were watching the exact same video?

**Leif:** They were. And they seemed to view the information in a biased way. For instance, 74% of conservatives said that the "military protestors" blocked pedestrians, but only 39% of conservatives said that the "abortion protestors" blocked pedestrians. The results for liberals were nearly the opposite: 76% said that "abortion protestors" blocked pedestrians, while only 45% of liberals said that the "military protestors" did so.

**Rosa:** Wow, that's kind of amazing how biased we can be in interpreting what we see and saying what we think the facts are about what we observed.

**Leif:** It is! It turns out, too, that people are willing to pay a monetary price to avoid seeing information and viewpoints that threaten their own political identity. In an article published in the *Journal of Experimental Social Psychology* in 2017, Jeremy Frimer and his co-authors report the results of a study in which they gave participants the option of reading controversial political opinions with which they disagreed for a $10 payment or, instead, the option to read political opinions with which they already agreed for a $7 payment. And guess what? About two-thirds of the study participants opted to read the material that aligned with their own political identity expressing opinions with which they already agreed. So, they accepted less money than they could have gotten, but by doing that they avoided having to interact with the identity-threatening material. Both Democrats and Republicans were equally likely to avoid the contrary information.

**Rosa:** It seems like they were motivated by unconscious drives to stay on their own political team's side and to keep viewing their team as the best team, since the economic incentive would have motivated them in the other direction.

**Leif:** Right. Here's another study that supports the idea that partisan identity – staying affiliated with a certain political party as part of a person's sense of self – matters more to people than policy content. Psychologist Geoffrey Cohen published results of a study (2003) showing that party labels win out over policy content in determining what people think of a proposal. He took a characteristically liberal policy for large welfare benefits and a characteristically conservative policy for small welfare benefits, and he presented them to a group of liberals and conservatives, who then indicated support for the policy that aligned, as predicted, with their liberal leanings or conservative leanings. But then he took a second group of people and presented the same proposals to them, but this time with labels on the policies as either having Democratic or Republican support, with the labels switched (so that, according to the labels on the policies, Democrats supposedly backed the conservative policy, and Republicans supposedly supported the liberal policy). It turned out that the study participants who were liberals supported the conservative proposal when they were told that it had Democratic support, and the study participants who were conservatives supported the liberal proposal when they were told that it had Republican support.

**Rosa:** When the study participants thought *their own political party* supported a policy, conservatives supported a generous welfare policy and liberals supported a meager welfare policy. That's wild that the label is what made the difference.

**Leif:** Yes, and subjects insisted that they had evaluated the policies on their merits and not on their partisanship, seeming to be completely unaware of the way their opinions of the proposals were driven by their sense of political identity and their desire to remain on their own team.

**Rosa:** Tell me how you see these kinds of studies as undermining human free will.

**Leif:** If our political beliefs tend to be motivated by wanting to stay a part of a team that is a core part of our sense of identity, rather than by interest in discovering the truth, then that means that what we say and do in social settings and in the political realm is also driven by these needs and desires. For example, we'd probably like to believe that we reason as we do about political matters by our own free thought rather than because we are driven by unconscious identity-preserving forces. And we'd probably like to believe that when we cast our votes in elections, we do so of our own free will. But these studies indicate that we are being driven by unconscious needs and biases that serve to preserve our sense of identity and belonging.

**Rosa:** This is all really interesting, and it has given me a lot to think about. Thanks for this conversation today, Leif.

**Leif:** I've enjoyed talking about it with you, Rosa! Next time we get together I'm going to describe some neuroscientific experiments that I think undermine free will.

**Rosa:** Sounds good, and I will read up on that topic beforehand. See you then.

# Day 10: Does Neuroscience Show that We Do Not Have Free Will?

**Abstract**

*Today Leif describes experiments begun in the 1980s by neurobiologist Benjamin Libet, who recorded brain activity called a "readiness potential," which he took to be an initiation of wrist flexing prior to a subject's report of having made a decision to flex, purportedly showing that decisions are made unconsciously and we only later become aware of our decisions, falsely believing that we made them for conscious reasons. Leif cites the work of scientists Daniel Wegener, Michael Gazzaniga, and Robert Sapolsky, as well, in support of the view that free will is an illusion of control over what we do, showing us to be somewhat pathetic in our sense of ourselves as agents who have conscious control. Rosa describes reasons for thinking that these experiments and theorizing do not undermine free will.*

**Rosa:** Hey, Leif! How are things going for you?

**Leif:** My linear algebra exam didn't go so well yesterday, but otherwise things are good. You?

**Rosa:** Just finished a painting I like, and I've been staying well despite my roommate being ill.

**Leif:** I'm glad! I'm ready to talk about more scientific evidence against free will today.

**Rosa:** Me, too. I read a short book by philosopher Alfred Mele over the past couple of days, which will help me back

DOI: 10.4324/9781003684268-13

up my position that science has not actually disproved free will.

**Leif:** I think science has undermined free will! Let's get started. You must have read about the influential work of neurobiologist Benjamin Libet (1985, 2004).

**Rosa:** Yes. I read that Libet started research in the early 1980s using a method for timing his subjects' conscious experiences that could be correlated with measurable activity in the brain.

**Leif:** Right. Participants in Libet's studies sat watching a spot revolve quickly around a clock face, and they were instructed to flex their wrist, whenever they felt like it, and to report where the spot was on the clock when they first were consciously aware of their urge or intention to flex their wrist. Readings of their brain activity were taken from their scalp while they did this, using electroencephalogram (EEG) technology measuring electrical conductivity. The participants flexed their wrists at least forty times during each testing session, and Libet also took readings from their wrist, so he could tell when the flexing muscle motion began.

**Rosa:** Got it.

**Leif:** When Libet repeated reminders to his test subjects to be spontaneous and not plan their wrist flexes, he found EEG results that looked like readings seen in prior studies of brain activity thought to be generated by areas at the front of the brain that precede and prepare intentional movements (brain activity called "readiness potential"), starting more than half a second – 550 milliseconds – before the wrist muscle started moving. The time of the participants' reported first awareness of their urge or intention or decision to flex their wrist was, on average, about one fifth of a second – 200 milliseconds – prior to the burst of muscle activity occurring when the wrist muscle started moving.

**Rosa:** This makes it look like activity starts in the brain – neurological activity that is leading up to the wrist flex – *before* the test subjects consciously feel an urge or intention to flex their wrist. And since the "readiness potential" is seen before the conscious awareness of any urge or decision to flex the wrist, the conclusion seems to be that it's really the brain starting the movement and unconsciously making the "decision" to flex, leading to the wrist movement – it's not the person's conscious intention or decision to flex that really causes the wrist flexing action.

**Leif:** Yes, exactly. Libet's interpretation of his experimental results was that the subject's decision to *flex my wrist now* was made when the readiness potential began, about a half second before the muscle motion began, whereas subjects' average time of first reported consciousness of an urge or decision to *flex my wrist now* was about a fifth of a second before the muscle motion began. And so Libet thought that his subjects were not really freely deciding to flex their wrists when they did – rather, the decision was made unconsciously first, leading to the muscle movement. Libet thought that, in order for free will to be involved in bringing about bodily actions, the decisions causing those bodily actions need to be made consciously. So he took his experiments to show that participants were not flexing their wrists of their own free will.

**Rosa:** Couldn't this apply only to his test subjects and not to the rest of us in ordinary life?

**Leif:** Well, I see why you ask, but I think you'd need to explain why that we be so. You'd need to set out the grounds for which you would dismiss Libet's results or take them only to apply narrowly. Libet did, indeed, take his results to generalize: he suggested that we never really make decisions consciously and so it's not free will that produces bodily actions. He did think, though, that once we become aware of our intention to do something, after the readiness potential has begun, we might have about a tenth of a second to veto the unconsciously formed intention to do that thing.

**Rosa:** Interesting. Experimental data is open to interpretation. I think it is unclear what that early recorded brain activity corresponds to or what it represents – I mean the electrical activity that's going on prior to the test participants' report of consciously intending to flex their wrist. Couldn't that early electrical activity just be them envisioning possibilities, or unconsciously thinking about whether or not to flex the wrist now? Alfred Mele points out that maybe what's happening in the brain when the rise in electrical activity starts is some process that might or might not lead to a decision to flex the wrist now. Libet's results do not rule that out, since Libet measured the rise in brain activity only for cases leading to wrist muscle movements. So, we don't know if that rise in brain activity also occurs in cases not leading to wrist flexes.

**Leif:** I think Libet's results show that conscious decisions do not actually cause bodily actions, despite people's mistaken impression that they do. Our actions begin earlier than we are aware of, as measured by the brain activity in his studies. So free will is an illusion.

**Rosa:** But here is another problem Mele raises against Libet's generalization from his experiments. In the wrist flexing studies, the subjects are told to be spontaneous, and they don't have any particular reasons for flexing when they do and not at other times instead. There's nothing big riding on what they do, like whether or not they get a good job. By contrast, the kind of decisions we usually think of as involving free will, at least many of them, have stakes attached to them, and we weigh pros and cons of making alternative decisions, such as in the case of whether to stay with my boyfriend or end the relationship, or the decision about what major academic field to study while in college. Maybe in cases like these more significant ones, when we consciously reason about what to decide, we do make our decisions consciously rather than unconsciously.

**Leif:** But I don't think you've given persuasive reasons for thinking that happens.

**Rosa:** Maybe so, but we're now in the midst of your turn to prove that we don't have free will. I'm not convinced that Libet's studies show that.

**Leif:** Okay, well, there is more scientific evidence. In an article, "Unconscious Determinants of Free Decisions in the Human Brain" (2008), a team of researchers including neuroscientist Chun Siong Soon used functional magnetic resonance imaging (fMRI) to assess brain activity by measuring changes in blood flow in areas of the brain. In the study, participants were instructed to press one or the other of two buttons, while their brain activity was measured. There was no penalty or reward associated with either button. The participants were told simply to choose, repeatedly. The scientists reported their discovery that two brain regions encoded with high accuracy, up to ten seconds *prior to* a conscious decision about which button to push, whether the participant was about to push the right button or the left button.

**Rosa:** I read about this study, and the accuracy that they said was "high" was only about 60%. If the correlation between the neural information they observed and the particular button pushing is there only 60% of the time, that leaves 40% of the time when it's not there, and that's not a whole lot higher than a 50–50 correlation by chance.

**Leif:** But it's still higher than chance.

**Rosa:** Yes, but there are other problems. That brain activity measured by blood flow might not be an unconscious decision to press a certain button. It might just be an unconscious inclination. And, like in the Libet studies, as you pointed out, there was nothing riding on the matter – who cares whether they push the right or left button? It doesn't matter. Their picking one or the other was arbitrary, and that's not like the cases I've talked about when I've consciously thought a lot about whether or not to do something, like to move to a different city or to start or end a significant relationship.

**Leif:** There's more. Here's another experiment, this one using depth electrodes placed on the surface of the brain or a bit beneath the surface, eliminating the need for the electricity measured by EEG to travel through the skull. In the journal *Neuron* (2011), the researchers, including neurosurgeon Itzhak Fried, conveyed their ability to predict with about 80% accuracy the timing of subjects' reported beliefs about when they first became aware of an urge to press a key. The predictions were made from readings taken from neurons in the supplementary motor area of the brain (involved in the production of bodily actions), and the predictions were made about 700 milliseconds in advance. Eighty percent of the time, the researchers measured changes in brain activity 700 milliseconds ahead that predicted within a few hundred milliseconds when subjects reported feeling a conscious urge to push the key. I think it's reasonable to believe that this result helps to show that unconscious brain activity has already determined when someone will press the key before they become aware of an urge to press it.

**Rosa:** Eighty percent accuracy in predicting the timing of subject reports doesn't sound like determinism to me. Also, isn't it really hard to be precise about when you first feel an urge or form an intention to press a key?

**Leif:** I'm making a cumulative case here, taking into account several studies that seem to have complimentary results in favor of my position. Other scientists also have been impressed by the evidence of Libet's studies and other studies like the ones I've described. The psychologist Daniel Wegener, for instance, in his book, *The Illusion of Conscious Will*, argues that free will is an illusion because our conscious intentions to act in certain ways are never among the causes for why we act as we do. He backs up his position by citing Libet's experiments along with other evidence.

**Rosa:** Like what?

**Leif:** Wegener cites evidence about automatic actions and movements people make without being aware of them. For example, tiny movements have been recorded in a person's hand going in sync with a metronome they hear, without the subjects being aware that they are making any movements at all. Also, it has been found that sometimes trained people who help patients with conditions that cause communication difficulties, such as cerebral palsy – these are trained people who are supposed to facilitate the patient in typing on keys to communicate the patient's own thoughts – in fact are unknowingly controlling the keys themselves, since what they end up typing was something known to the facilitator but not the patient. There are also cases in which certain kinds of brain damage can lead to automatic behaviors, such as the patient's automatically putting on a pair of eyeglasses whenever they are handed the glasses, even if they're already wearing a pair of glasses. Wegener's conclusion is that our actions are uniformly automatic and bypass our conscious will. We only have an illusion of consciously being in charge of what we do.

**Rosa:** Well, Mele shows quite well the errors in Wegener's reasoning. Even if some actions are automatic and not caused by conscious intentions, it does not follow that *all* our actions are that way. Mele points out that Wegener fails to support the assumption that all actions are caused basically in the same way. So when Wegener reasons that some of what we do is not caused even partly by conscious intentions, therefore none of what we do is caused even partly by conscious intentions, Wegener is making an unsupported leap.

**Leif:** Interesting point.

**Rosa:** In fact, Mele goes further than that. He gives evidence for thinking it is true that some of our actions are caused at least partly by conscious intentions. Mele cites studies in which people do better in following through on what they say they're going to do and in meeting goals, such as taking up an exercise routine or finishing a task like writing a resume by the end of the day, when they consciously make a plan for when they are going to do it, forming "implementation intentions." This is why, when volunteers come to your door in support of getting out the vote, they ask you if you have made a specific plan for voting. If you do have an implementation intention or specific plan for where and when you're going to vote, you're more likely to actually go cast your vote in an election.

**Leif:** Still, there are many other scientists who are more impressed than you and Mele are with the neuroscientific evidence. Neuroscientist Michael Gazzaniga, in his book, *Who's In Charge? Free Will and the Science of the Brain* (2011), writes that, since we have no access to the unconscious processing that really explains what we do, we *invent* explanations for the behavior we've observed ourselves doing, but these are just rationalizations after the fact, not true causal explanations. Gazzaniga is impressed by experiments including the ones I described by Libet and by Soon and his team.

**Rosa:** I see.

**Leif:** There's also a well-known neuroscientist and primatologist at Stanford University, a winner of the MacArthur genius grant, named Robert Sapolsky. In his book,

*Determined: A Science of Life Without Free Will* (2023), Sapolsky contends that free will is a myth and that every living organism is just a biological machine, since factors like our hormones, childhood, life experiences, and our circumstances all meld together to produce actions that we only feel were ours to choose freely. Sapolsky says that neuroscience has gotten really good at understanding what particular parts of the brain do, using techniques like neuroimaging and transcranial magnetic stimulation, and has also gotten very good at understanding how variants of particular genes interact with particular enzymes and what happens in neurons – and we have good reason to expect science to keep getting better at accurately explaining why we behave in the ways we do.

**Rosa:** What else does Sapolsky say?

**Leif:** He says that, because we are biological machines and free will is a myth, we should stop thinking that we get what we deserve in life, and we should try to change how we think about things. We should realize, for instance, that who we end up loving is really only a matter of how our parents raised us and what sort of olfactory receptor genes we have in our noses and how much we like the loved person's scent. Sapolsky says it's impossible to act freely of everything that has come before. Our idea that we are the "captains of our ships" is just a delusion.

**Rosa:** It sounds like maybe he thinks a free act would have to be completely uncaused.

**Leif:** It's hard to say exactly, but he does write this: "Show me a neuron being a causeless cause in this total sense... Show me a neuron (or brain) whose generation of a behavior is independent of the sum of its biological past, and for the purposes of this book, you've demonstrated free will" (2023, p. 15).

**Rosa:** I get the impression it would have been useful if Sapolsky had given more analysis in his book of what free will amounts to, along with more attention to philosophical work on the issue of whether or not free will is compatible with causal determinism and with causal indeterminism.

**Leif:** I could agree to that.

**Rosa:** Since I knew from our last conversation that you were going to cite neuroscientists today, I went looking last night for other expert opinions. I found that Kevin Mitchell, a neuroscientist and geneticist at Trinity College Dublin, in his book, *Free Agents: How Evolution Gave Us Free Will* (Princeton University Press, 2023), writes this: "Despite the headlines proclaiming the death of free will, it remains stubbornly alive and well. Nothing in philosophy or physics or neuroscience or genetics or psychology or neurology or any other science undermines the idea that we do have the capacity for conscious, rational control of our actions…The broad idea of moral responsibility is…unaffected by discoveries from science that are revealing the neural and cognitive underpinnings of rational control" (Mitchell, 2023, pp. 282–283).

**Leif:** So you're pointing out that there is not uniform agreement on what neuroscience shows about free will?

**Rosa:** Yes. And I haven't been convinced by what you've said today.

**Leif:** I understand. You've raised objections that I'll think about some more.

**Rosa:** Would you want to recap your case against free will, describing your arguments from the past several times we have gotten together?

**Leif:** Sure! First, I argued that, if a perfect, omniscient, omnipotent God exists, then we do not have free will. I do not myself believe that God exists. But, as you pointed out at the start, millions and even billions of people do believe that God exists. I argued that, when religious believers think about God's attributes, including perfect divine knowledge and perfect divine omnipotence, they should conclude that we do not have free will. That's because human free will cannot be successfully reconciled with divine omniscience, and it cannot be successfully reconciled with divine omnipotence, either.

**Rosa:** You did give me a lot to think about that day.

**Leif:** Next, I argued that causal determinism rules out free will. We cannot have free will, if causal determinism is true. In other words, we can have free will only if causal determinism is false. But there is good reason to think that causal determinism is true. So, we do not actually have free will.

**Rosa:** I wasn't convinced that you had proved that causal determinism is true.

**Leif:** Well, then I argued in this way: suppose causal determinism is false. Suppose what is true, instead, is causal indeterminism – the denial of the thesis of causal determinism. I argued that free will is not consistent with causal indeterminism either. You described different theories of free will that depict it as being incompatible with determinism but as being compatible with indeterminism. I argued that those theories do not give people the control that we think of as needed for free will. Those accounts depict human beings as subject to luck and randomness.

**Rosa:** I found the indeterminist theories we talked about more plausible and attractive than you did.

**Leif:** I remember. Next, I described arguments in social psychology that show that we do not have free will, including experiments on the bystander effect, Stanley Milgram's studies on obedience to authority (involving study participants' administering electric shocks), and Philip Zimbardo's Stanford prison experiment. I also pointed to more recent studies on politically motivated reasoning, showing how much we are driven by unconscious desires to belong and to protect our sense of political identity.

**Rosa:** Those were all really interesting to talk about.

**Leif:** Finally, my fifth argument was that a body of neuroscientific evidence undercuts human free will. I cited several scientific studies, including Benjamin Libet's famous experiments, showing that the belief that we initiate our own behavior for conscious reasons is incorrect and free will is an illusion.

**Rosa:** I was skeptical about what the studies show, but I still think they're worth looking into more, which I will do. Anything more you want to say today about free will?

**Leif:** I think that's it for now. Maybe we should take some time on our own for a few days and get together next week, after we've had some time to mull over our various days of conversations.

**Rosa:** That sounds great, Leif. Next time, we could try to summarize what we've talked about over the span of these conversations, and then figure out where we agree and disagree. We could also talk about some implications of where we stand on free will. How about I get back in touch about a time?

**Leif:** Sounds terrific! See you sometime next week, Rosa.

# Conclusion

# Day 11: Summing Up

**Abstract**

*As their final conversation about free will opens, we find that Rosa and Leif both have come to soften their initial opinions. They each explain how their thinking has changed. Their conversation turns to practical implications of what they have discussed, including implications for the criminal justice system and for our attitudes about government assistance to citizens in need. Although Rosa and Leif remain in disagreement over whether or not God exists, they agree that monotheistic religious traditions do not provide clear, noncontroversial answers regarding human free will. They also agree that, whether we have free will in a libertarian sense or (only) in a compatibilist sense, we can genuinely love each other, and human life can have meaning. An additional point of agreement they find is that, if we do not have libertarian free will, then some of us may need an adjustment in our self-image in relationship to the future. Still, even if we do not have libertarian free will, we do not need to stop conceiving of ourselves as agents, as beings who act.*

**Rosa:** Hi Leif! I'm so glad we've had these conversations. Why don't I start out just quickly reminding us of the reasons I gave for thinking that we have free will, since it's been a while.

**Leif:** Sure. I'm glad, too, Rosa – it's been fun to talk! I'll briefly recap my arguments against free will, as well.

DOI: 10.4324/9781003684268-15

**Rosa:** Sounds great. I argued that we have free will for these reasons. One, God holds us responsible for what we choose to do, and God's gift of free will explains why there is suffering and evil in the world. So being a theist provides reasons for believing in human free will. Two, we are morally responsible to each other for our actions, which shows that we have free will, since free will is required for moral responsibility in the sense of our deserving praise or blame for what we do. Three, our humanity involves our having free will, or maybe more specifically the fact that we human beings are creative shows that we have free will, and the fact that human life has meaning shows that we have free will. Four, we love each other (and our pets and projects and other things), and love requires free will, so we must have free will. Five, we can tell from the inside, in our own experience, that we have free will.

**Leif:** Thanks for that summary, Rosa. Here's my case against free will. One, if there is a perfect God who is omniscient and omnipotent, then we do not have free will, because divine knowledge of all truths rules out our ever being able to act in any way other than the way in which God knows we will act, and if God has all power and is in charge of everything that happens in the universe, that does not leave any room for any of us to be in charge of the way things go. So, if the billions of people who believe in God are correct, then human beings have no free will. Two, causal determinism appears to be true, and it rules out free will. It rules out free will because it rules out our ever being ultimately in charge of what we do and because it rules out our ever having genuine options available to us for how to act, since on determinism there is only one physically possible future at every moment. Three, even if causal determinism is false and instead causal indeterminism is true, we still don't have free will because indeterminism only introduces randomness into our decision-making processes and so into our behavior. Four, experiments in social psychology undermine our illusion of free will by showing that that we are driven to behave as we do by our situations, our social roles, and our unconscious desires to please others, to belong, and to preserve our sense of social and political identity. Five, neuroscientific experiments undercut free will by providing evidence that unconscious brain activity initiates our behavior prior to our having any awareness of what we supposedly "decide" to do.

**Rosa:** I've been glad to hear your different arguments against free will, Leif, especially since I had always just assumed that we do have it. I had taken it for granted, as long as I can remember. You know what, as I've thought back over everything we have talked about over the course of our conversations, I think I have come to soften my position somewhat.

**Leif:** Oh, really? Tell me about that – how have your views changed?

**Rosa:** Well, at the start of our conversations, I realize now that I was what you'd call a metaphysical libertarian: I believed that we do have free will, and that free will is incompatible with causal determinism but compatible with causal indeterminism, and that causal determinism is false. I recognize now that the libertarian sense of freedom is the one I originally had in mind. I thought of free will as requiring genuinely open options, where there are (sometimes) multiple physically possible futures in front of the free person, and it is up to them to choose which one to make the actual future. I envisioned my own future that way. And on the basis of my sense of myself in relation to my future, as well as on the basis of my religious beliefs and my beliefs about love, meaning, humanity's value and creativity, and moral responsibility, I believed that we do in fact have libertarian free will.

**Leif:** I see. Do you still stand firm in your belief that we have libertarian free will?

**Rosa:** After these days of talking with you and reading more on the topic, I now see it as more of an open question whether or not we have free will in the libertarian sense.

**Leif:** Why is that?

**Rosa:** It's partly because I do appreciate the worries you raised for libertarian accounts – maybe they just involve randomness. I'm not sure. Maybe we are not able to be "agent causes" at all, or maybe we are not able to be agent causes in a way that gives us control rather than it just being a matter of luck or randomness whether we agent-cause one decision or agent-cause a different decision. And maybe the presence of causal indeterminism in our decision-making processes would not allow us to settle the matter of which future becomes the actual one, as you alleged in response to the event-causal indeterminist accounts of free will.

**Leif:** I see.

**Rosa:** But my change in perspective is also partly because you've given me reasons to think that maybe causal determinism is true, after all. It's not that I've become convinced that causal determinism is, in fact, true. It's that I better recognize my limitations in knowing how the universe works. I realize that I'm not a scientist. How would I know for certain whether or not causal determinism is true? I realize now that, in the past, I had assumed that it was false, but I hadn't really thought about it much. Also, how would I know for certain whether or not there actually is causal indeterminism in our decision-making processes of a kind that gives us control over what we decide, amounting to free will? It seems like these are open empirical questions that I am just not qualified to answer on my own. And it sounds to me like even scientists disagree and do not have clear, unambiguous answers.

**Leif:** Yes, that sounds reasonable to me, Rosa. I recognize that I am not a scientist, either. I acknowledge that, as a non-expert, I don't have full insight into the way the universe runs. I don't really know a lot about the ways our brains work as we deliberate over what to do, either.

**Rosa:** Have your views about free will changed at all since we've been talking?

**Leif:** You know, I think I have also come to soften my initial opinion somewhat.

**Rosa:** How so?

**Leif:** When we started talking, I was adamantly against the idea that we have free will. I felt pretty certain that we have no such thing, and that "free will" is just an invention or illusion.

**Rosa:** What changed?

**Leif:** Well, now I know more about the various philosophical theories of free will. We talked about how philosophers have distinguished different ways we can go about giving an account of what free will amounts to. You raised some good objections to the scientific experiments we talked about, both in the realm of social psychology and in neuroscience. Maybe I too easily took on board, from reading about those experiments and reading those scientific writers, an inchoate idea that our having free will meant that we had to have immaterial souls, which I don't believe in, or it would have to involve something happening that is completely uncaused but that still somehow amounts to control of ours over what we do.

**Rosa:** I see.

**Leif:** I guess now it seems to me that it is possible – and maybe even likely – that we do have free will in at least some compatibilist sense. By that I mean, it seems reasonable to believe that we do sometimes act without coercion or constraint, and we are sometimes able to act freely as depicted, for instance, by Dana Kay Nelkin and by Susan Wolf – to act rightly from an appreciation of right reasons – or as depicted by Gary Watson or by Harry Frankfurt – to act from our values or to act from our hierarchically-aligned desires. I used to be dogmatically attached to the position that absolutely everything we do is initiated by unconscious brain activity without, for instance, our reflectively-formed desires about our desires being central to explaining our actions. Now I'm more open to the idea of our having free will as it is described on one or another of the compatibilist accounts of what free will is.

**Rosa:** Could you say more about what has led you to think that?

**Leif:** As we've talked and I've reflected on it more, I've found myself not convinced, in the end, that we always act automatically or only on unconscious impulses, and I'm not convinced that we are completely driven to do exactly what we do by our situations or by the chemicals in our brains. I don't doubt that we are heavily influenced by factors like wanting to obey authority, please others, conform to our social roles, and belong to a certain political group.

**Rosa:** I agree with you that we are influenced by those forces and pressures.

**Leif:** Still, none of the experiments we talked about do seem to prove conclusively that we never act for conscious reasons. They don't prove that we never intend to do what we do without being forced by unconscious impulses toward social conformity or identity preservation. Once I learn about the bystander effect, for instance, and once I learn about the human tendency to conform to obey authority, I can take conscious steps to counter those influences over me. Education helps me to free myself from unconscious biases and influences!

**Rosa:** I agree.

**Leif:** Now that I've thought about it more, it seems pretty clear to me that sometimes I can consciously intend to do something, like make a plan to meet a friend such as you at a certain time, and that I can carry out that intention as I had planned, of my own accord, for reasons of mine, whether those are my hierarchically-aligned desires, or my values, or as an expression of my ability to do the right thing for right reasons, such as to keep my word to you, my friend. Even in the face of the experiments in social psychology and neuroscience, it seems plausible to think that some of the mechanisms that lead to my action are reasons-responsive ones of my own, as John Martin Fischer's theory emphasizes. I recognize that I could be wrong, but before I was just completely dismissive of the idea that we can sometimes act freely.

**Rosa:** I get it. It sounds like neither of us is as entrenched in our beliefs as we were before. I agree with you about what you've just said. It seems to me that there are not any especially powerful arguments against the idea that we are able sometimes to act freely in a compatibilist sense. Neuroscience and social science have not shown, for instance, that we *never* act on reasons-responsive mechanisms, and it seems evident to me that we sometimes do. That type of free will is compatible with causal determinism, so we don't have to worry about scientists someday demonstrating that determinism is true – we can have free will even if they were to do that. Maybe you and I could turn now to talking about some practical implications of the topics we have talked about.

**Leif:** Sure. Do you have certain implications in mind?

**Rosa:** Yes. One question I have in mind is this. Should incompatibilists about free will and causal determinism who think that causal determinism is true – in other words, hard determinists, as you used to be – as well as incompatibilists who think that it is unclear whether or not determinism is true – someone like me, who now sees it as an open scientific question – advocate for changes to the criminal justice system? Likewise, should those who think that free will is incompatible with determinism and also incompatible with indeterminism, so we can't have it either way, advocate for criminal justice system changes?

**Leif:** I'd love to talk about that.

**Rosa:** And I have another question. Should incompatibilists of the sorts I just described advocate for changes in people's attitudes about government assistance to citizens who are in need, for instance, in need of food and housing assistance programs? I remember when you first brought up the topic of free will, you had been bothered by something your father had said on the phone. It sounded like maybe he was in effect blaming people who lack housing as being at fault for their situation since it resulted from their own free choices. And maybe he adopted his political positions in part based on that view of his about people's free will.

**Leif:** Yes, I remember that. Let's take your first question first, about our criminal justice system.

**Rosa:** Sure. What do you think?

**Leif:** I think hard determinists, free will skeptics, and incompatibilists like you – incompatibilists who see it as unsettled scientifically whether or not causal determinism is true – probably should advocate for significant changes to the criminal justice system.

**Rosa:** I agree. But tell me why you say that.

**Leif:** I say that because incompatibilists believe that compatibilist free will is not enough for anyone to genuinely deserve punishment or blame. And if no one *deserves* punishment for their wrong actions, then we really ought to think more carefully about how people are treated after they do something wrong, especially in prisons. I can understand the intuition that, if a person's life follows a straight and unforked road to prison from birth, then it's simply not fair for us to lock people up in prisons where they are treated harshly in a retributive way, as if this is justified by "tipping back the scales of justice," because it's not justified. If no one can ever help doing what they do in that it's the only thing that is physically possible for them to do at the time, then they can't help doing things that are wrong. Yet some people in existing prisons are treated horribly for all the rest of their lives or even are put to death.

**Rosa:** Yes, my main concern is fairness.

**Leif:** I understand.

**Rosa:** Since I think that libertarian free will is required for anyone to genuinely deserve punishment and blame, and since I think it is unreasonable to assert that we, in fact, have libertarian free will, given the limited state of our scientific knowledge and given the disagreement there is within the scientific community, I see now that my position should be that our retributive practices should stop and our retributive attitudes like anger and resentment should change. The ways we treat each other and the attitudes we take towards each other should be different. Incompatibilists, at least those who are not staunch libertarians, should hold that, *if* our practices of praising and blaming people, and punishing and rewarding people, are justified at all, they are not currently justified in the sense that people *deserve* those responses to their good and bad actions. Wrongdoers should not be harmed as if they deserve to be treated horribly. We should see criminal behavior as something to be managed, with more compassion. We still need to protect people in societies from criminal behavior, but our responses to it should be focused on effective methods of molding people's future behavior.

**Leif:** I see that those are implications of your views about free will. About your second question regarding government assistance: I think that hard determinists – and also people who are skeptical about the existence of free will because they think it is both incompatible with determinism and incompatible with indeterminism – should in common advocate widely for the position that people are not to blame for their need of help through governmental social assistance programs due to their own supposed "free" choices in a desert sense of blame.

**Rosa:** Yes, and incompatibilists like me – those of us who are unconvinced that we actually have libertarian free will and are unsure about its existence – we, too, should spread the message that at least we don't *know* that other people are to blame for their actions or for their need for social assistance, and they might well not be. So perhaps we ought to act with more humility, generosity, and compassion toward others.

**Leif:** That seems right to me, too, Rosa.

**Rosa:** I'm glad I've realized this: since I think libertarian free will is required for deserved blame and punishment, and I don't know whether or not people have libertarian free will, and I don't know whether or not causal determinism is true, I should not judge people harshly or think I know they deserve retribution when they act in wrong ways.

**Leif:** I understand.

**Rosa:** Maybe you could say more about compatibilists regarding these topics. What are some implications concerning criminal justice and social attitudes, from your perspective?

**Leif:** I think compatibilists about free will and determinism, and compatibilists about free will and indeterminism, do not necessarily have to think that our legal and social practices of imposing punishment should change, so long as they think it is reasonable to believe that we do have compatibilist free will and so long as punishments are within ethical limits such as upholding basic human rights. I think compatibilists can hold, too, that our common interpersonal moral responsibility practices and attitudes, like gratitude and resentment, can coherently and sensibly remain in place. As I said, through our conversations I've come to see it as reasonable to believe that, even in the face of pronouncements to the contrary from some theorists in the social sciences and in the neuroscientific community, people can at least sometimes act with compatibilist free will.

**Rosa:** I suppose someone could object by saying that we don't really *know* that we sometimes act on reasons-responsive mechanisms, for instance, as on Fischer's theory of free action. But we talked over reasons to doubt that the social scientific experiments and the neuroscientific ones we examined really show what some scientists have pronounced that they show and completely undermine free will.

**Leif:** Yes, we did. I also think that even compatibilists who believe in free will and who do think that punishment is sometimes deserved could oppose capital punishment and advocate for our criminal justice systems to be more humane. Just because someone thinks that blame and punishment are sometimes deserved doesn't mean that they approve of all forms that punishment and blame might take – of course, they can think that there are ethical principles governing how we treat all people that still apply.

**Rosa:** I see that.

**Leif:** You and I started from opposite sides on the question of free will. After these conversations, what are some other points of agreement and disagreement we have?

**Rosa:** Well, I still believe that God exists. You don't, right?

**Leif:** Right. Like I said, I think of myself as at least an agnostic and probably an atheist.

**Rosa:** But maybe we can find some agreement about the connection between free will and God's existence. At the start, I thought my belief in God gave good reasons to think that we have free will, and I was thinking of free will in a libertarian way.

**Leif:** I remember.

**Rosa:** But now I think that monotheistic religious traditions do not provide clear, noncontroversial answers regarding human free will. As our conversations and more reading on the topic helped me to see, different philosophers and theologians within theistic traditions have differing ways of conceiving of human freedom and the nature and implications of divine omnipotence and divine omniscience.

**Leif:** That I do agree with. What do you think about some of your other initial arguments?

**Rosa:** I've thought about them a lot. What I have come to think now is that, whether we have free will in a libertarian sense or (only) in a compatibilist sense, we can genuinely love each other, and human life can have meaning.

**Leif:** Why do you say that?

**Rosa:** You convinced me of it when we talked about love and meaning, once I learned about different accounts of what love and meaning are. I came to see how we could still love each other and our projects and pets and the natural world, and how we can make meaningful lives, by way of control over our actions by reasons, in a way that is compatible with determinism. I still love my brother, and that love is real, whether I have libertarian free will or not. And there is meaning in my life, coming in part from my being actively engaged in my relationships with good people and in my artistic projects, regardless of whether or not there is ever more than one physically possible future. Neither love nor meaning requires libertarian free will.

**Leif:** That seems right to me, too.

**Rosa:** If it turns out that we do not have libertarian free will, then I do think that an adjustment is needed in the self-image of any of us who envision ourselves as agents who face an open future – an open future in the sense of our sometimes having more than one future in front of us that is physically possible and that is up to us to make actual.

**Leif:** Agreed. Some people might need to change their self-image. Even if we do not have libertarian free will, though, I think we do not need to stop conceiving of ourselves as agents, as beings who act.

**Rosa:** I get that now. I don't have to think that causal determinism's truth would imply that I am just a falling leaf in the wind or that I should just lie in bed and be passive. I am capable of forming intentions and plans, capable of having goals and a sense of purpose, and capable of putting my plans into action. That's what it is to be an agent rather than an object. You and I aren't just things like blankets or rocks. We are beings who can think, and plan, and act intentionally. We don't necessarily need to have free will to be able to do those things, as agents.

**Leif:** Well, that seems like a good place to stop, don't you think? Thanks so much for these talks, Rosa. They've helped me a lot to clarify my thinking about free will.

**Rosa:** Thanks to you, too, Leif! Maybe some other time we could start a series of conversations about the ethics of abortion, or about climate change, or about personal identity.

**Leif:** Sounds great, but for now I need a little break from philosophy!

**Rosa:** (Laughs.) Yes, me too. See you around, Leif.

# Annotated Bibliography

## *Day 1: Does Religion Show that We Have Free Will?*

Augustine, St. (c. 400–426/1991–2019). *The Works of Saint Augustine. A Translation for the 21st Century*, 46 vols., John E. Rotelle et al. (eds.). New York: New City Press.
Augustine (354–430) defends the doctrine of original sin, the idea that human beings inherit a fallen state making us prone to sin and in need of God's grace for salvation. He supports the idea that we are free beings who are accountable to God and who will be judged by God for our actions. Augustine's most famous works are *Confessions* and *City of God*.

Aquinas, St. Thomas (c. 1270/1981). *Summa Theologica*, trans. Fathers of the English Dominican Province. Westminster: Christian Classics.
Aquinas (1225–1274) is among the most eminent of the world's theologians. This comprehensive work of theology explores the nature of God and demonstrates Aquinas' belief in human accountability and divine judgment.

Ekstrom, Laura (2021). *God, Suffering, and the Value of Free Will*. Oxford: Oxford University Press.
This work is an in-depth examination of free will in relation to God and evil, in which the author questions the worth of free will and argues against the rationality of believing in hell, both of which are concerns voiced by Leif in the dialogues.

Ekstrom, Laura. (2023). *Evil and Theodicy*. Cambridge Elements in the Problems of God. Cambridge: Cambridge University Press.
This short, accessible book assesses proposed theological justifications for the suffering in our world, including the contrast, appreciation, character development, and free will theodicies expressed by Rosa in the dialogue.

McBrayer, Justin and Daniel Howard-Snyder, eds. (2013). *The Blackwell Companion to the Problem of Evil*. Malden, MA: Wiley-Blackwell.
This volume offers a wide array of articles on religious belief in relation to evil, including arguments from evil for atheism, along with historical and contemporary responses to those arguments, including theodicies from Jewish, Hindu, Christian, and Islamic perspectives, some of which appeal to human free will.

Peterson, Michael (2022). *Monotheism, Suffering, and Evil.* Cambridge Elements in Religion and Monotheism. Cambridge: Cambridge University Press.
Peterson discusses in this concise work how Judaism, Christianity, and Islam respond to the problem of evil and recommends a best line of response to the argument from gratuitous suffering.

Pew Research Center (2012). "The Global Religious Landscape," December 18, 2012. www.pewresearch.org/religion/2012/12/18/global-religious-landscape-exec/
This comprehensive demographic study of more than 230 countries conducted by the Pew Research Center's Forum on Religion & Public Life reveals that more than eight in ten people worldwide identify with a religious group. It estimates that there are 5.8 billion religiously affiliated people around the globe, amounting to 84% of the 2010 world population of 6.9 billion.

Pew Research Center (2021). "Few Americans Blame God or Say Faith Has Been Shaken by Pandemic, Other Tragedies," section 2, "Views on the Afterlife," November 23, 2021. www.pewresearch.org/religion/2021/11/23/views-on-the-afterlife/
This study provides data on the percentages of American adults who believe in heaven and in hell, broken down by specific religious affiliation, political affiliation, and region of the U.S.

Plantinga, Alvin (1974). *God, Freedom, and Evil.* Grand Rapids, MI: Wm. B. Eerdmans Publishing.
Plantinga relies on free will in his defense against the claim that it is inconsistent to believe in the reality of God along with acknowledging that there is evil in the world.

Schellenberg, John (2004). "The Atheist's Free Will Offense." *International Journal for Philosophy of Religion* 56: 1–15.
Schellenberg argues that, if we have free will, then God does not exist.

Speak, Daniel (2015). *The Problem of Evil.* Malden, MA: Polity Press.
Speak provides an engaging overview of the main lines of reasoning in the debate over God, free will, and suffering, and he argues for the defensibility of theistic belief in the face of evil.

Swinburne, Richard (1998). *Providence and the Problem of Evil.* Oxford: Oxford University Press.
Swinburne's response to the question of what justifies God in allowing the world's evils rests in part on both the existence of, and the great positive value of, human free will. Swinburne supports the idea that free will is a gift to humanity from God, an idea espoused by Rosa in the dialogue.

Talbott, Thomas (2014). *The Inescapable Love of God*, 2nd edition. Eugene, Oregon: Cascade Books.
Talbott argues that divine love rules out there being an eternal hell endured by any human beings, a view expressed by Leif in the dialogue.

Van Inwagen, Peter (2006). *The Problem of Evil*. Oxford: Clarendon Press.
Van Inwagen offers, in response to arguments from evil, a defensive story that he says is true for all we know, one that shows God's existence to be reconcilable with the evils of this world. Van Inwagen's defense relies heavily on the existence and positive value of free will.

## *Day 2: Does Interpersonal Moral Responsibility Show that We Have Free Will?*

Coates, D. Justin and Neal Tognazzini, eds. (2012). *Blame: Its Nature and Norms*. New York: Oxford University Press.
The volume is a collection of thought-provoking essays on questions about the human practice of blame, including what it is to blame, what blame expresses, whether blame is a vice, and what standing a person needs to have to blame another person.

Ekstrom, Laura (2000). *Free Will: A Philosophical Study*. Boulder, CO: Westview Press.
As part of the larger project in this book, Ekstrom argues that moral responsibility requires the ability to do otherwise not in a conditional sense described by Leif in the dialogue but instead in the sense that the person has genuine options for acting at the time – multiple pathways into the future such that it is up to her which becomes actual.

Fischer, John Martin (2012). *Deep Control: Essays on Free Will and Value*. Oxford: Oxford University Press.
This book is a collection of Fischer's influential essays on moral responsibility, which address topics including sourcehood, guidance control, and arguments for the compatibility of determinism and moral responsibility.

Fischer, John Martin and Mark Ravizza (1998). *Responsibility and Control: A Theory of Moral Responsibility*. Cambridge: Cambridge University Press.
Fischer and Ravizza set out an influential account of moral responsibility according to which it requires what they call guidance control, which is compatible with causal determinism, and does not require what they call regulative control, which would be ruled out by causal determinism.

Hobbes, Thomas (1654/1999). "Of Liberty and Necessity," in ed. Vere Chappell, *Hobbes and Bramhall on Liberty and Necessity*. Cambridge: Cambridge University Press, pp. 15–42.
Hobbes gives a conditional analysis of the ability to do otherwise needed for free will.

Hume, David (1740/1978). *A Treatise of Human Nature*, eds. L. A. Selby-Bigge and P. H. Nidditch, 2nd edition. Oxford: Oxford University Press.
Hume understands freedom as involving a lack of external impediments to a person's will, and he argues that it is not only compatible with determinism but requires determinism.

Lehrer, Keith (1968). "Cans without Ifs." *Analysis* 29: 29–32.
Lehrer critically assesses the conditional analysis of the ability to do otherwise connected to free will.

Nelkin, Dana Kay (2011). *Making Sense of Freedom and Responsibility*. Oxford: Oxford University Press.
Nelkin's theory of moral responsibility centers on the ability to do the right thing for the right reasons. She argues that praise and blame are compatible with determinism.

Pereboom, Derk (2001). *Living Without Free Will*. Cambridge: Cambridge University Press.
Pereboom understands free will as the control over action needed for moral responsibility in the desert sense, and he argues that this free will is incompatible with determinism. Since it is unlikely on our best evidence, he argues, that we have free will, Pereboom advocates for skepticism about free will and moral responsibility. He explores the implications of such skepticism.

Strawson, Peter (1963). "Freedom and Resentment." *Proceedings of the British Academy* 48: 187–211.
This widely discussed essay sets out Strawson's expressivist theory of our human practices of praising, blaming, punishing, and rewarding ourselves and others.

van Inwagen, Peter (1986). *An Essay on Free Will*. Oxford: Oxford University Press.
Van Inwagen defends the view that free will requires the categorical ability to do otherwise than act as one does at a time, and he argues that no one has this ability if causal determinism is true. Moral responsibility requires such free will, van Inwagen contends, so moral responsibility is incompatible with determinism. Taking moral responsibility to be indubitable, van Inwagen concludes that determinism should be rejected.

Wolf, Susan (1990). *Freedom Within Reason*. Oxford: Oxford University Press.
Wolf sets out an account of the freedom needed for moral responsibility as the ability to act in accordance with the True and the Good. She explores and critically assesses alternative accounts.

## *Day 3: Does Our Humanity Show that We Have Free Will?*

Fischer, John Martin (2009). "Stories and the Meaning of Life." *Philosophical Exchange* 39 (no. 1): 2–16.
In this paper, Fischer argues that, when a person acts freely, she writes a sentence in the story of her life, which gives her life the shape of a narrative and thereby provides her life with a unique sort of meaning and value. Meaning in life thus involves making a statement, rather than making a difference. Making a statement can be achieved by a person who has guidance control over her action, which is compatible with determinism, and does not require regulative control, which would require the categorical ability to do otherwise.

Kane, Robert (1996). *The Significance of Free Will.* Oxford: Oxford University Press.

In this book, Kane presents and defends a theory of free will; along the way, he connects free will to human dignity and creativity, as Rosa does in the dialogues.

Pereboom, Derk (2014). *Free Will, Agency, and Meaning in Life.* Oxford: Oxford University Press.

Pereboom defends the idea that human life can be meaningful, and we can retain distinctive human features and relationships, even if we lack free will.

Singer, Peter (1986). "All Animals Are Equal," in ed. Peter Singer, *Applied Ethics.* Oxford: Oxford University Press, pp. 215–228.

Singer argues that there is not a good reason to think that human beings are all inherently more valuable or worthy of respect and consideration than are other non-human animals (not, for instance, our intelligence or our alleged free will), contrary to Rosa's ideas expressed in the dialogue.

Tolstoy, Leo (1882/2008). "My Confession," in eds. E. D. Klemke and Steven Kahn, *The Meaning of Life: A Reader,* 3rd ed. New York: Oxford University Press.

Tolstoy describes a period of his life in which he found no sense of purpose in his projects, art, science, family, literature, learning, or physical comforts. He concludes that meaning in life lies in choosing religious faith and devotion to God.

Wolf, Susan (2012). *Meaning in Life and Why It Matters.* Princeton, NJ: Princeton University Press.

Wolf provides an account of a meaningful human life as one in which a person loves what is worth loving.

## *Day 4: Does Love Show that We Have Free Will?*

Frankfurt, Harry (1971). "Freedom of the Will and the Concept of a Person." *Journal of Philosophy* 68, 5–20.

Frankfurt suggests that the ability of persons to rise above their desires and form desires about those desires is key to understanding human freedom. He defends the view that a person acts freely when she acts as she wants to act and, crucially, she also wants to want to act in that way, so that there is hierarchical alignment between her first-level and her second-level desires.

Frankfurt, Harry (2004). *The Reasons of Love.* Princeton, NJ: Princeton University Press.

Frankfurt argues that love involves a volitional necessity in which people cannot help caring about who or what they care about.

Pereboom, Derk (2001). *Living Without Free Will.* Cambridge: Cambridge University Press.

Pereboom argues that, although we are neither free nor morally responsible, much of what matters in life remains intact; along the way, he argues that love does not require free will.

Stump, Eleonore (2006). "Love, By All Accounts." *Proceedings and Addresses of the American Philosophical Association* 80 (2): 25–43.

Stump defends an account deriving from the work of St. Thomas Aquinas according to which love is a desire for union with the beloved appropriate to the station between them and a desire for the flourishing of the beloved.

van Inwagen, Peter (1988). "The Magnitude, Duration, and Distribution of Evil: A Theodicy." *Philosophical Topics* 16: 161–187.
In this work, as part of his treatment of questions concerning God and evil, van Inwagen argues that love of God and love of other people requires libertarian free will.

Watson, Gary (1975). "Free Agency." *Journal of Philosophy* 72, 205–220.
Watson criticizes Frankfurt's account of human free action and suggests, instead, that we act freely when we act from our values.

## *Day 5: Does the Feeling of Freely Choosing Show that We Have Free Will?*

Horgan, Terrance (2015). "Injecting the Phenomenology of Agency into the Free Will Debate," in ed. David Shoemaker, *Oxford Studies in Agency and Responsibility*, vol. 3. Oxford: Oxford University Press, pp. 34–61.
Horgan explores the features of our experience of ourselves as freely performing some of our actions.

Kant, Immanuel (1785/1981). *Grounding for the Metaphysics of Morals*, trans. James Ellington. Indianapolis, IN: Hackett.
Kant contends that we can act only "under the idea of freedom" and for that very reason we are free from a practical point of view, an idea Rosa appeals to in the dialogues.

Lehrer, Keith (1960). "Can We Know That We Have Free Will by Introspection?" *The Journal of Philosophy* 57: 145–157.
Lehrer argues that our sense of freedom is not necessarily a reliable guide to whether or not we have free will and that our understanding of free will can be grounded in empirical observations rather than in introspection.

Nahmias, Eddy, Stephen Morris, Thomas Nadelhoffer, and Jason Turner (2004). "The Phenomenology of Free Will." *Journal of Consciousness Studies* 11: 162–179.
In this paper, the authors present disputes about the phenomenology of free will, survey psychological research on relevant experiences, and describe their own studies. They conclude that compatibilist descriptions of the phenomenology of free will are better supported by the data than libertarian descriptions.

Nelkin, Dana Kay (2011). *Making Sense of Freedom and Responsibility*. Oxford: Oxford University Press.
As part of this book's broader aims, Nelkin closely analyzes our sense of freedom and suggests that it crucially involves a desire to be an explanatory nexus, such that what we do is explained by our deliberation over reasons.

Reid, Thomas (1788/1983). *Essays on the Active Powers of Man*, in eds. R. E. Beanblossom and K. Lehrer, *Inquiry and Essays*. Indianapolis: Hackett.

Reid suggests that human beings have a natural conviction of their own ability to act freely, and he contends that this conviction throws the burden of proof to those who allege that we do not have such a power.

## *Day 6: Does God's Existence Show that We Do Not Have Free Will?*

Aquinas, St. Thomas (c. 1270/1981). *Summa Theologica*, trans. Fathers of the English Dominican Province. Westminster: Christian Classics.
Aquinas views God as existing atemporally, and he seeks to explain how God's knowledge can be reconciled with human freedom.

Boethius (c. 500–524/1962). *The Consolation of Philosophy* (Prose III, Poem III, Prose IV, Prose VI), trans. Richard Green. Macmillan Publishing Company.
Boethius (480–524) was a Roman scholar who, like Aquinas, depicted God's knowledge as atemporal or outside of time.

Origen of Alexandria (c. 246/2002). *Commentary on the Epistle to the Romans, Books 6-10*, trans. Thomas P. Scheck. Washington, D.C.: The Catholic University of America Press.
Some philosophers attribute the "dependence solution" to the problem of reconciling divine omniscience and human free will to Origen, as discussed by the characters in the dialogue.

Pike, Nelson (1965). "Divine Omniscience and Voluntary Action." *The Philosophical Review* 74 (no.1): 27–46.
Pike argues that divine foreknowledge is logically incompatible with human free will. Therefore, either an omniscient God exists or we have free will, but both cannot be true at once.

## *Day 7: Does Causal Determinism Show that We Do Not Have Free Will?*

Daniel, Dennett (1984). *Elbow Room: The Varieties of Free Will Worth Wanting*. Cambridge, MA: MIT Press.
Dennett argues that incompatibilist analogies for human life on the assumption of determinism – such as that it would be like being in prison or being a puppet or a plaything – are inaccurate and overblown "bugbears." He defends an account of human control over action that is compatible with determinism.

Ekstrom, Laura (2000). *Free Will: A Philosophical Study*. New York: Routledge.
Ekstrom uses the paradise island scenario, discussed by Leif in the dialogue, to motivate incompatibilism. She defends other arguments for incompatibilism and critically evaluates arguments for compatibilism concerning free will and causal determinism. Ekstrom also sets out a particular indeterminist model of free will.

Mele, Alfred (2006). *Free Will and Luck*. New York: Oxford University Press.

As an aspect of his larger project in this book, Mele presents a "zygote argument," similar to the argument for incompatibilism discussed by Leif and Rosa in the dialogues, which aims to shift the burden to compatibilists to show what relevant difference there is between, on the one hand, an agent's being manipulated (or originally designed as a zygote by a goddess) to do some particular action at a time and, on the other hand, being (merely) causally determined to do it.

Pereboom, Derk (2001). *Living Without Free Will*. Cambridge: Cambridge University Press.

As part of his case for incompatibilism, Pereboom presents a four-case argument starting from an example in which there is covert manipulation of a person by direct stimulation of her brain, leading through other cases involving programing and indoctrination, to a case of a person's being causally determined to do what she does. By relying on the principle of treating like cases alike, Pereboom concludes that in none of the cases is the person morally responsible for what she does, due to the fact that the sources of the person's actions can be traced back in their entirety to originating conditions that were completely beyond her control.

van Inwagen, Peter (1986). *An Essay on Free Will*. Oxford: Oxford University Press.

In this book, van Inwagen presents his famous consequence argument and no choice principles in support of the incompatibility of free will and causal determinism. Thus, van Inwagen thinks, if causal determinism is true, then we do not have free will.

## *Day 8: Does Causal Indeterminism Show that We Do Not Have Free Will?*

Chisholm, Roderick (1976). "The Agent as Cause," in eds. Myles Brand and Douglas Walton, *Action Theory*. Dordrecht, Holland: D. Reidel.

Chisholm argues that persons have free will only if they are sometimes unmoved movers who are able to agent cause, as substances, their volitions. Chisholm thus holds that not all causes of effects are events.

Church, Ian and Robert Hartman, eds. (2019). *The Routledge Handbook of the Philosophy and Psychology of Luck*. New York: Routledge.

The volume contains an array of articles on the nature of luck and its implications for human agency and free will, which may interest readers wanting to explore these matters more fully.

Ekstrom, Laura (2019). "Toward a Plausible Event-Causal Indeterminist Account of Free Will." *Synthese* 196 (no. 1): 127–144.

Ekstrom explores difficulties for event-causal indeterminist accounts of free will and defends one such account against objections.

Franklin, Christopher (2018). *A Minimal Libertarianism: Free Will and the Promise of Reduction*. Oxford: Oxford University Press.

Franklin argues against the contention that free will libertarians must appeal to substance agent-causation of the sort posited by Chisholm, and he defends an indeterminist model of free will on which the causal role of the agent reduces to states and events involving the agent.

Ginet, Carl (1990). *On Action*. Cambridge: Cambridge University Press.
Ginet defends a non-causal libertarian account of free will.

Griffith, Meghan (2017). "Agent-Causal Libertarianism," in eds. Neil Levy, Meghan Griffith, and Kevin Timpe, *The Routledge Companion to Free Will*. New York: Routledge, pp. 72–85.
Griffith defends an agent-causal libertarian account of human free will.

Kane, Robert (1996). *The Significance of Free Will*. New York: Oxford University Press.
Kane argues for incompatibilism, explores why free will matters, and articulates a particular event-causal libertarian account of freedom centering on the notion of "self-forming actions."

Mele, Alfred (2006). *Free Will and Luck*. New York: Oxford University Press.
Mele commits neither to compatibilism nor incompatibilism in this book, but he develops two different conceptions of free will – one for incompatibilists and the other for compatibilists. Mele suggests that luck poses problems for all believers in free will, and he offers solutions to those problems on behalf of both compatibilists and incompatibilists.

O'Connor, Timothy (2000). *Persons and Causes*. New York: Oxford University Press.
O'Connor develops and defends a particular agent-causal libertarian theory of free will.

Strawson, Galen (1998). "Luck Swallows Everything." *Times Literary Supplement*, June 26.
Strawson argues that being free and ultimately morally responsible requires one to be *causa sui* (the cause of oneself, at least in certain crucial mental respects), but no one can be the ultimate cause of oneself, so freedom and ultimate moral responsibility are logically impossible.

## *Day 9: Does Social Psychology Show that We Do Not Have Free Will?*

Cohen, Geoffrey (2003). "Party Over Policy: The Dominating Impact of Group Influence on Political Beliefs." *Journal of Personality and Social Psychology* 85: 808–822.
Cohen's study supports the claim that partisan identity matters more to people than policy content. Study subjects insisted that they evaluated the study's proposals on their merits and not on their partisanship, seeming to be completely unaware of the way their opinions were driven by their sense of political identity.

Darley, John M., & Latane, Bibb. (1968). "Bystander Intervention in Emergencies: Diffusion of responsibility." *Journal of Personality and Social Psychology* 8: 377–383.
Darley and Latane report the results of a study testing the hypothesis that, the larger a group of potential helpers a person thinks they are in, the less quickly and the less likely they are to help a victim in an emergency.

Frimer, Jeremy, Linda Skitka, and Matt Moytl (2017). "Liberals and Conservatives Are Similarly Motivated to Avoid Exposure to One Another's Opinions." *Journal of Experimental Social Psychology* 72: 1–12.
Frimer and his colleagues describe their study indicating that subjects were motivated by unconscious drives to stay on their own political team's side and to continue viewing their team as the best team, since the economic incentives involved in the study would have motivated them in a different direction.

Kahan, Dan, David Hoffman, Donald Braman, Danieli Evans, and Jeffrey Rachlinski (2012). "They Saw a Protest: Cognitive Illiberalism and the Speech-Conduct Distinction." *Stanford Law Review* 64: 851–906.
The authors report the results of their study demonstrating how people's loyalty to "their side" in politics shapes how they interpret information. In this study, all subjects watched a video of a protest but were told different things about the kind of protest being observed.

Milgram, Stanley (1974). *Obedience to Authority*. New York: Harper & Row.
At Yale University, psychologist Stanley Milgram conducted a series of studies designed to test the ways people behaved when put in experimental conditions in which they were told by an authority to do something that would seem to go against personal conscience. In his studies, participants were assigned to roles as either "teachers" or "learners," with the former led to believe that they were administering electric shocks to the latter for wrong answers.

Zimbardo, Philip (n.d.). "Stanford Prison Experiment," www.prisonexp.org/
Zimbardo, a Stanford University psychologist, conducted a famous study in which participants were assigned to roles as prisoners or guards. The aim of the study was to examine the effects of situational variables on participants' reactions and behaviors.

## *Day 10: Does Neuroscience Show that We Do Not Have Free Will?*

Balaguer, Mark (2010). *Free Will as an Open Scientific Problem*. Cambridge, MA: MIT Press.
Balaguer argues that, since a certain event-causal libertarian account of free will is best, the question of whether or not we have free will boils down to a straightforward empirical question about whether certain neural events in our heads are causally undetermined in a specific way. Since there is no good evidence as to whether or not the relevant neural events are undetermined in the way that is required, Balaguer concludes, the question of whether human beings have is an open scientific question.

Fried, Itzhak, Roy Mukamel, and Gabriel Kreiman (2011). "Internally Generated Preactivation of Single Neurons in Human Medial Frontal Cortex Predicts Volition." *Neuron* 69: 548–562.
Fried and his research team report their ability to predict with about 80% accuracy the timing of subjects' reported beliefs about when they first became aware of an urge to press a key, indicating that unconscious brain activity has already determined when someone will press the key before they become aware of an urge to press it.

Gazzaniga, Michael (2011). *Who's in Charge? Free Will and the Science of the Brain*. New York: Harper Collins.

Gazzaniga defends the view that, since we have no access to the unconscious processing that really explains our behavior, we invent explanations for what we've done, but these are merely rationalizations after the fact and not true causal explanations.

Libet, Benjamin (1985). "Unconscious Cerebral Initiative and the Role of Conscious Will in Voluntary Action." *Behavioral and Brain Sciences* 8: 529–566.

This article reports Libet's study of unconscious cerebral initiative in voluntary action by measuring brain activity prior to finger or wrist flexing.

Libet, Benjamin (2004). *Mind Time*. Cambridge, MA: Harvard University Press.

In this book, Libet gives an account of his experiments that reveal a delay between the unconscious processes that he takes to initiate our behaviors and our conscious experiences, showing, he concludes, that free will is an illusion.

Mele, Alfred (2014). *Free: Why Science Hasn't Disproved Free Will*. New York: Oxford University Press.

This short, readable book rebuts the work of many scientists who claim that there is no such thing as free will.

Mitchell, Kevin (2023). *Free Agents: How Evolution Gave Us Free Will*. Princeton: Princeton University Press.

Mitchell, a geneticist and neuroscientist, argues that free will is not a mystical, nebulous property that is divinely implanted, but instead is an evolved biological function that enables human beings to consciously, rationally control our actions. Mitchell concludes that nothing in the sciences undermines human moral responsibility.

Sapolsky, Robert (2023). *Determined: A Science of Life Without Free Will*. New York: Penguin Press.

Sapolsky argues that, in light of evidence in the sciences, belief in human free will is only a pleasant fantasy. He suggests that, although it may be difficult for us to give up this fantasy, accepting the non-reality of free will can lead to a more humane world.

Soon, Chun Siong, Marcel Brass, Hans-Jochen Heinze, and John-Dylan Haynes (2008). "Unconscious Determinants of Free Decisions in the Human Brain." *Nature Neuroscience* 11: 543–545.

Neuroscientist Soon and his colleagues used functional magnetic resonance imaging (fMRI) to assess brain activity by measuring changes in blood flow in areas of the brain. They report that two brain regions encoded with high accuracy, up to ten seconds prior to a conscious decision about which button to push, whether the participant was about to push the right button or the left button.

Wegener, Daniel (2002). *The Illusion of Conscious Will*. Cambridge, MA: MIT Press.

Wegener argues that free will is an illusion because our conscious intentions to act in certain ways are never among the causes for why we act as we do.

# Index

Printed in the United States
by Baker & Taylor Publisher Services